Advanced Maths Essentials
Statistics 1 for AQA

Welcome to Advanced Maths Essentials: Statistics 1 for AQA. This book will help you to improve your examination performance by focusing on all the essential skills you will need in your AQA Statistics 1 exam. It has been divided by chapter into the main topics that need to be studied. Each chapter has then been divided by sub-headings, and the description below each sub-heading gives the AQA specification for that aspect of the topic.

The book contains scores of worked examples, each with clearly set-out steps to help solve the problem. You can then apply the steps to solve the Skills Check questions in the book and past exam questions at the end of each chapter. If you feel you need extra practice on any topic, you can try the Skills Check Extra exercises and even more exam questions on the accompanying CD-ROM. At the back of this book there is a sample exam-style paper for the without-coursework option and on the CD-ROM there is one for the with-coursework option, to help you test yourself before the big day.

Some of the questions in the book have a ⊚ symbol next to them. These questions have a PowerPoint® solution (on the CD-ROM) that guides you through suggested steps in solving the problem and setting out your answer clearly.

Using the CD-ROM

To use the accompanying CD-ROM simply put the disc in your CD-ROM drive, and the menu should appear automatically. If it doesn't automatically run on your PC:

1. Select the My Computer icon on your desktop.
2. Select the CD-ROM drive icon.
3. Select Open.
4. Select statistics1_for _aqa.exe.

If you don't have PowerPoint® on your computer you can download PowerPoint 2003 Viewer®. This will allow you to view and print the presentations. Download the viewer from http://www.microsoft.com

Pearson Education Limited
Edinburgh Gate
Harlow
Essex
CM20 2JE
England
www.longman.co.uk

First published 2005
ISBN   0 582 836905

Design by Ken Vail Graphic Design

Cover design by Raven Design

Typeset by Tech-Set, Gateshead

Printed in the U.K. by Scotprint, Haddington

The publisher's policy is to use paper manufactured from sustainable forests.

The publisher wishes to draw attention to the Single-User Licence Agreement at the back of the book.
Please read this agreement carefully before installing and using the CD-ROM.

We are grateful for permission from AQA to reproduce past exam questions. All such questions have a reference in the margin. AQA can accept no responsibility whatsoever for accuracy of any solutions or answers to these questions.

Please note that the following AQA (AEB) and AQA (NEAB) questions used:- p.14, p.41, p.53, p.54, p.64, p.65, are NOT from the live examinations for the current specification. For GCE Advanced Level subjects, new specifications were introduced in 2001.

Every effort has been made to ensure that the structure and level of sample question papers matches the current specification requirements and that solutions are accurate. However, the publisher can accept no responsibility whatsoever for accuracy of any solutions or answers to these questions. Any such solutions or answers may not necessarily constitute all possible solutions.

1 Numerical measures

1.1 Standard deviation and variance

Standard deviation and variance calculated on ungrouped and grouped data.

The **standard deviation** is a measure of the average deviation of the values from their mean.

The formula depends on whether the values represent a **population** or a **sample**.

Standard deviation and variance of a population

For n **population** values, $x_1, x_2, \ldots, x_n$, the **standard deviation**, denoted by σ, is given by

$$\sigma = \sqrt{\frac{1}{n}\sum_{i=1}^{n}(x_i - \mu)^2} = \sqrt{\frac{\sum x^2}{n} - \mu^2}$$

where $\mu = \dfrac{\sum x}{n}$ is the population mean.

The (population) standard deviation is **never negative**, as it is based on a sum of squares.

The **population variance**, σ^2, is the square of the population standard deviation.

Example 1.1 The number of pupils in a school's seven classes are

$$19, 21, 23, 25, 24, 20, 22$$

Calculate the standard deviation and variance of the school's class sizes.

Step 1: Calculate the mean as 'sum of values' ÷ 'number of values'.

Population mean, $\mu = \dfrac{19 + 21 + \ldots + 22}{7} = \dfrac{154}{7} = 22$ pupils

Step 2: Calculate the 'sum of squares of the values'.

Sum of squares, $\sum x^2 = 19^2 + 21^2 + \ldots + 22^2 = 3416$

Step 3: Substitute the calculated values into the formula used for calculation.

Population standard deviation, $\sigma = \sqrt{\dfrac{3416}{7} - 22^2} = \sqrt{488 - 484}$
$$= \sqrt{4} = 2 \text{ pupils}$$

Population variance, $\sigma^2 = 2^2 = 4$

Standard deviation and variance of a sample

For n **sample** values, $x_1, x_2, \ldots, x_n$, the **standard deviation**, denoted by s, is given by

$$s = \sqrt{\frac{1}{n-1}\sum(x-\bar{x})^2} = \sqrt{\frac{1}{n-1}\left(\sum x^2 - \frac{(\sum x)^2}{n}\right)}$$

where $\bar{x} = \dfrac{\sum x}{n}$ is the sample mean.

The (sample) standard deviation is **never negative**, as it is based on a sum of squares.

The **sample variance**, s^2, is the square of the sample standard deviation.

The reason for the divisor $(n-1)$, rather than n, in the formula for the sample standard deviation is to ensure that, in repeated sampling from a population, the average value of the sample variance, s^2, will tend to the population variance, σ^2.

Note:
The first expression for s is the definition; the second is an equivalent form for calculation purposes. Also $\sum x^2 \neq (\sum x)^2$.

Recall:
$\bar{x}$ is read as 'x-bar'.
$\bar{x}$ and μ require the same calculation.

Note:
An expression for s^2 is provided in the examination. From this you can find the first expression for s.

Note:
This is mentioned again in Section 5.1.

Example 1.2 The values below are the weights, in grams, of the contents of a sample of nine tins of tuna.

$$187, 185, 188, 184, 185, 186, 189, 187, 188$$

Calculate the standard deviation of these weights.

Note:
As $\bar{x} = \dfrac{1679}{9} \approx 186.5556$, the use of the definition formula introduces computational errors.

Step 1: Calculate the 'sum of values'.

Sum of weights, $\sum x = 1679$

Note:
Scientific and graphical calculators may be used in the examination and may have a sample standard deviation function (s, s_{n-1} or σ_{n-1}) and a population/sample mean function ($\bar{x}$).

Step 2: Calculate the 'sum of squares of the values'.

Sum of squares of weights, $\sum x^2 = 313\,249$

Step 3: Substitute the calculated values into the formula used for calculation.

Standard deviation $= \sqrt{\dfrac{1}{8}\left(313\,249 - \dfrac{1679^2}{9}\right)} = \sqrt{\dfrac{22.222}{8}}$

$$= 1.67 \text{ grams (3 s.f.)}$$

For a **frequency distribution**, the sample standard deviation is given by

$$s = \sqrt{\frac{1}{\sum f - 1}\sum f(x-\bar{x})^2} = \sqrt{\frac{1}{\sum f - 1}\left(\sum fx^2 - \frac{(\sum fx)^2}{\sum f}\right)}$$

where f denotes the frequency of the value x, and $\bar{x} = \dfrac{\sum fx}{\sum f}$ is the sample mean.

Recall:
A **frequency distribution** is a list of (integer) values together with their corresponding frequencies.

Note:
The first expression for s is the definition; the second is an equivalent form for calculation purposes. These expressions are **not** provided in the examination.

Note:
$\sum fx^2 \neq (\sum fx)^2$.

Example 1.3 The number of paper clips in each of a sample of 110 boxes is recorded with the following results.

Number of paper clips	99	100	101	102	103	104
Number of boxes	9	46	42	8	2	3

Recall:
These are **discrete data** arising from **counting** (the number of paper clips).

Calculate the standard deviation of these data.

Note:
In fx^2, only the x is squared.

Step 1: Identify the x-values (and so the f-values).

Step 2: Use a table to record values for $\sum fx$ and $\sum fx^2$.

Clips (x)	Boxes (f)	fx	fx^2
99	9	81	88 209
100	46	4600	460 000
101	42	4242	428 442
102	8	816	83 232
103	2	206	21 218
104	3	312	32 448
Total	110	11 067	1 113 549

Note:
Scientific and graphical calculators may be used in the examination and most have a standard deviation function (s, s_{n-1} or σ_{n-1}) that caters for a sample frequency distribution.

Tip:
Get to know all the relevant functions on your calculator by checking the answers to these examples.

Step 3: Substitute the calculated values into the formula used for calculation.

Standard deviation, $s = \sqrt{\dfrac{1}{109}\left(1\,113\,549 - \dfrac{11\,067^2}{110}\right)}$

$\quad\quad = 0.996$ paper clips (3 s.f.)

Recall:
The sample mean is given by
$\bar{x} = \dfrac{11\,067}{110}$ or 100.6 paper clips
(1 d.p.).

For a **grouped frequency distribution**, an estimate of the sample standard deviation is given by

$$s = \sqrt{\frac{1}{\sum f - 1}\sum f(x - \bar{x})^2} = \sqrt{\frac{1}{\sum f - 1}\left(\sum fx^2 - \frac{(\sum fx)^2}{\sum f}\right)}$$

where f denotes the class frequency, x the value of the corresponding class midpoint, and $\bar{x} = \dfrac{\sum fx}{\sum f}$ is the sample mean.

Recall:
A **grouped frequency distribution** is a list of (continuous) classes together with their corresponding frequencies.

Note:
These are the same formulae as those for a frequency distribution and are **not** provided in the examination.

Example 1.4 The lengths, to the nearest centimetre, of a sample of 100 steel rods are summarised in the following table.

Length (cm)	Number of rods
110 to 119	8
120 to 129	26
130 to 134	21
135 to 139	19
140 to 149	12
150 to 159	10
160 to 179	4

Recall:
These are **continuous data** arising from **measuring** (the lengths of the rods).

Calculate an estimate of the standard deviation of the lengths of these rods.

Step 1: Find the class midpoints (x).

Step 2: Use a table to record values for $\sum fx$ and $\sum fx^2$.

Length (cm)	Number of rods (f)	Class midpoint (x)	fx	fx^2
110 to 119	8	114.5	916.0	104 882.00
120 to 129	26	124.5	3237.0	403 006.50
130 to 134	21	132.0	2772.0	365 904.00
135 to 139	19	137.0	2603.0	356 611.00
140 to 149	12	144.5	1734.0	250 563.00
150 to 159	10	154.5	1545.0	238 702.50
160 to 179	4	169.5	678.0	114 921.00
Total	100		13 485.0	1 834 590.00

Note:
As individual values are not known, only an estimate of the standard deviation can be calculated.

Note:
In fx^2, only the x is squared.

Note:
Most scientific and graphical calculators have a standard deviation function (s, s_{n-1} or σ_{n-1}) that caters for a sample grouped frequency distribution, although it may be necessary to enter the class midpoints.

Step 3: Substitute the calculated values into the formula used for calculation.

Estimate of the standard deviation,

$$s = \sqrt{\frac{1}{99}\left(1\,834\,590 - \frac{13\,485^2}{100}\right)} = 12.8 \text{ cm (3 s.f.)}$$

Recall:
An estimate of the sample mean is given by $\bar{x} = \dfrac{13\,485}{100}$ or 135 cm (3 s.f.).

SKILLS CHECK **1A: Standard deviation and variance**

1 Included in the contents of a desk drawer were six unopened boxes of drawing pins. The number of pins found in each box was as follows:

    51   49   54   48   50   48

Calculate the mean and standard deviation for the number of drawing pins in these boxes.

2 The values shown are the times, in minutes, taken by a random sample of 10 trainee plumbers to change a washer on a tap.

     5.9   5.4   8.6   3.8   5.4   6.3 7.1   2.9   4.4   6.2

Calculate the mean and standard deviation of this sample of times.

3 The marks obtained by a class of 20 students in a statistics test are as follows:

    48   31   25   51   22   42   38   36   58   45
    28   11   34   35   47   27   41   26   23   52

Calculate the mean and variance of the marks obtained by these 20 students.

 4 As part of her statistics coursework, Sarah records her bus journey times to school on a random sample of 15 days. The times, in minutes, are as follows:

    15.7   14.8   12.3   18.6   15.8   13.4   14.7   17.6   16.3
    15.4   13.8   16.7   15.1   14.6   12.6

  a Use these data to estimate the mean and standard deviation of Sarah's bus journey time to school.

  b Explain why Sarah did not simply record her bus journey times over a 3-week period.

5 The number of customers entering a small butcher's shop is recorded for a random sample of 2-minute periods with the following results.

Number of customers	0	1	2	3	4	5	6
Number of 2-minute periods	12	35	56	37	21	11	8

Calculate the mean and standard deviation for the number of customers per 2-minute interval entering this butcher's shop.

 6 A pair of six-sided dice, each with the numbers 1, 2, 3, 3, 4 and 5 on its faces, are rolled 150 times. On each roll, the score is the sum of the numbers on the uppermost faces of the two dice. The results are shown in the table.

Score	2	3	4	5	6	7	8	9	10
Number of occurrences	5	10	19	22	34	27	18	12	3

Calculate the mean, variance and standard deviation of the score obtained on a roll of these two dice.

7 The weights, to the nearest gram, of a random sample of adult male dormice in summer are shown in the table.

Weight	10 to 11	12 to 13	14 to 15	16 to 17	18 to 19	20 to 21	22 to 23	24 to 25
Number of dormice	2	7	13	29	21	15	10	3

Estimate the mean and standard deviation of the weight of the adult male dormice in summer.

 8 The lengths, in centimetres, of a random sample of 75 cheese-topped baguettes are summarised as follows.

Length	31–	33–	35–	36–	37–	38–	39–	41–	43–45
Number of baguettes	5	8	9	13	16	11	7	5	1

Calculate estimates of the mean, variance and standard deviation of cheese-topped baguettes.

9 The lengths, in centimetres to the nearest millimetre, of a random sample of 2-metre strips of plastic conduit are summarised in the table.

Length	Number of strips
198.0 to 198.4	3
198.5 to 198.9	4
199.0 to 199.4	6
199.5 to 199.9	15
200.0 to 200.4	56
200.5 to 200.9	49
201.0 to 201.4	38
201.5 to 201.9	32
202.0 to 202.4	21
202.5 to 202.9	15
203.0 to 203.4	6
203.5 to 203.9	2
204.0 to 205.0	3

Use the mean and standard deviation functions on your calculator to find estimates of the mean, standard deviation and variance of lengths of 2-metre strips of plastic conduit.

SKILLS CHECK **1A EXTRA** is on the CD

1.2 Linear scaling

Linear scaling.

If $y = ax + b$, where a and b are constants, then

$$\bar{y} = a\bar{x} + b \quad \text{and} \quad s_y^2 = a^2 s_x^2$$

Note:
If $a > 0$ then $s_y = as_x$.
If $a < 0$ then $s_y = -as_x$.

Note:
These formulae are **not**
supplied in the examination.

Example 1.5 The times, x seconds, in excess of 30 seconds, that a sample of buses wait at a bus stop, have a mean of 18 and a standard deviation of 15. Find the mean and standard deviation of the times that this sample of buses wait at the stop.

Step 1: Determine the form of the linear relationship between y and x.

Denoting the times, in seconds, that buses wait at the stop by y, then $y = x + 30$ so the mean, $\bar{y}$, is given by

$$\bar{y} = 18 + 30 = 48 \text{ seconds}$$

Step 2: Substitute for a and b in the formulae.

and the standard deviation, s_y, is given by

$$s_y = s_x = 15 \text{ seconds}$$

Note:
Here $a = 1$ so $s_y = s_x$.

Example 1.6 A particular type of steel rod has a nominal length of 5 metres. The deviations, x millimetres, from this nominal length for a sample of these steel rods have a mean of 110 and a standard deviation of 48. Find, in metres, the mean and standard deviation of the lengths of this sample of steel rods.

Step 1: Determine the form of the linear relationship between y and x.

Denoting the lengths, in metres, of the steel rods by y, then $y = \dfrac{x}{1000} + 5$ so the mean, $\bar{y}$, is given by

Step 2: Substitute for a and b in the formulae.

$$\bar{y} = \frac{110}{1000} + 5 = 5.11 \text{ metres}$$

and the standard deviation, s_y, is given by

$$s_y = \frac{48}{1000} = 0.048 \text{ metres}$$

> **Recall:**
> There are 1000 mm in a metre.

Example 1.7 For a particular period during 1904, the mean and standard deviation of the maximum daily temperatures, $x\,°F$, are 54.6 and 2.3. Convert these to °C using $y = \dfrac{5}{9}x - \dfrac{160}{9}$.

Step 1: Substitute for a and b in the formulae.

In °C, the mean is $\dfrac{5}{9} \times 54.6 - \dfrac{160}{9} = \dfrac{113}{9} = 12.6$ (3 s.f.).

In °C, the standard deviation is $\dfrac{5}{9} \times 2.3 = 1.28$ (3 s.f.).

SKILLS CHECK **1B: Linear scaling**

1 The measurements x have a mean of 30 and a variance of 16. Given that $y = 100 + 10x$, find the mean and variance of the measurements y.

2 The random variable R has a mean of -15 and a variance of 4. Given that $t = 20 - 2r$, find the mean and standard deviation of the values of t.

3 The weight, in grams in excess of one kilogram, of sugar in a sample of bags has a mean of 10 and a standard deviation of 4. Calculate, in kilograms, the mean and standard deviation of sugar in a bag.

4 Suresh, an amateur timekeeper, starts his stopwatch exactly 4 hours after the start of a cross-country exercise. He then records all finishing times from his stopwatch in minutes. The mean and standard deviation of his recorded times for all participants on the exercise are 78.6 and 14.7, respectively. Calculate, in hours, the mean and standard deviation of the participants' times to complete the exercise.

 5 Whilst on a fact-finding trip on the continent, Charles records the difference in price, in € from €5, of 1-litre bottles of a particular make of virgin olive oil in a sample of French outlets. The differences recorded are as follows.

    0.33 -0.67 1.01   0.51 -0.25 0.75 -0.49

  a Calculate the mean and standard deviation of this sample.

  b Given an exchange rate of €1.56 to the £, calculate, in £s to the nearest 1p, the mean and standard deviation of a 1-litre bottle of this particular make of virgin olive oil in France.

SKILLS CHECK **1B EXTRA is on the CD**

Mode, median, mean, range, interquartile range and standard deviation.

The **mode** (or modal value) is the value that occurs most frequently.

Advantages
- Easy to find
- If it exists, it is an observed value
- May be used for non-numeric data

Disadvantages
- May not be unique or may not exist
- May be unrepresentative
- Difficult to estimate from grouped data
- Basis for minimal further analysis

The **median** is the middle value of a string of **ordered** values.

When the number of values, n, is **odd**, the median is the $\left(\dfrac{n+1}{2}\right)$th ordered value.

When the number of values, n, is **even**, the median is the **average** of the $\left(\dfrac{n}{2}\right)$th and the $\left(\dfrac{n}{2}+1\right)$th ordered values.

> **Recall:**
> **Ordered** implies that the values are rearranged in order of magnitude with the smallest value to the left and the largest value to the right.

Advantages
- Can sometimes be found when only partial data is available
- Useful when there are **outliers**

Disadvantages
- Difficult to estimate from grouped data
- Basis for only some further analysis

> **Note:**
> **Outliers** are unusually large or unusually small values, when compared to other values in the data set.

The (arithmetic) **mean**, or **average**, is the sum of the values divided by the number of values.

Advantages
- Takes account of every value
- Basis for further analysis

Disadvantages
- In a small sample, may be unduly affected by outliers or incorrect values

As a general rule, statisticians favour the use of the mean as a **measure of average** but if it is not appropriate they use the median. Rarely do they use the mode.

The **range** is the difference between the highest and lowest values.

Advantages
- Easy to calculate

Disadvantages
- Depends only on extreme values
- Only appropriate for small data sets

The **interquartile range** (IQR) is the difference between the upper and lower quartiles.
The **lower quartile**, Q_1, is the median of the ordered values to the left of the median.
The **upper quartile**, Q_3, is the median of the ordered values to the right of the median.

> **Recall:**
> The median is also the second quartile, Q_2.

Advantages
- Can sometimes be found when only partial data is available
- Useful when there are outliers

Disadvantages
- Difficult to estimate from grouped data
- Basis for only some further analysis

The **standard deviation** is a measure of the average deviation of the values from their mean.

Advantages	*Disadvantages*
• Takes account of every value • Basis for further analysis	• In a small sample, may be unduly affected by outliers or incorrect values • Difficult to calculate from large data sets

As a general rule, statisticians favour the use of the standard deviation as a **measure of spread** but if it is not appropriate they use the IQR or, for small sets ($n < 10$) with no outliers, the range.

Example 1.8 For each of the following data sets:

   i   find the mode, median and mean

   ii find the range, interquartile range and standard deviation

   iii state, with reasons, for each of **i** and **ii**, which, if any, is the most appropriate.

   a The numbers of mixed mushrooms in a sample of 15 cartons:

      10, 9, 10, 11, 8, 9, 13, 10, 10, 15, 12, 12, 30, 14, 13

   b The times, in minutes, taken by a sample of 12 children to complete a puzzle:

      3.43, 3.92, 4.24, 3.98, 3.74, 4.03, 4.11, 3.76, 3.55, 3.65, 3.89, 4.10

Step 1: Order the values.
a i   Ordering the 15 values gives

      8, 9, 9, 10, 10, 10, 10, 11, 12, 12, 13, 13, 14, 15, 30

Step 2: Identify the value with the greatest frequency.

  Mode = 10 mushrooms

Step 3: Identify the middle value.

  Median = 11 mushrooms

Step 4: Calculate 'sum of values' ÷ 'number of values'.

$$\bar{x} = \frac{186}{15} = 12.4 \text{ mushrooms}$$

> **Recall:**
> The mean may be a value that cannot actually occur.

Step 5: Identify the highest and lowest values, then calculate the difference.

ii Range = 30 − 8 = 22 mushrooms

Step 6: Identify the median of (i) values to the right of the median, and (ii) values to the left of the median, then calculate the difference.

  IQR = 13 − 10

      = 3 mushrooms

> **Tip:**
> To find Q_3 and Q_1, split the data at the median, then find the median of each part.

Step 7: Calculate 'sum of squares of values', then substitute the values into the computational formula.

$$s = \sqrt{\frac{1}{14}\left(2694 - \frac{(186)^2}{15}\right)} = 5.26 \text{ mushrooms (3 s.f.)}$$

> **Tip:**
> Use your calculator to find $\bar{x}$ and s directly, taking care with data entry.

iii The median, as it is an observable value and is not affected by the relatively large value of 30.

    The IQR, as it is also not affected by the relatively large value of 30 and is usually quoted with the median.

Step 1: Order the values. **b i** Ordering the 12 values gives

$$3.43, 3.55, 3.65, 3.74, 3.76, 3.89, 3.93, 3.98, 4.03, 4.10,$$
$$4.14, 4.24$$

There is no mode.

Step 2: Identify the value with the greatest frequency.
Step 3: Calculate 'average of middle two values'.

$$\text{Median} = \frac{3.89 + 3.93}{2} = 3.91 \text{ minutes}$$

Step 4: Calculate 'sum of values' ÷ 'number of values'.

$$\bar{x} = \frac{46.44}{12} = 3.87 \text{ minutes}$$

Step 5: Identify the highest and lowest values, then calculate the difference.

ii Range $= 4.24 - 3.43 = 0.81$ minutes

Step 6: Identify the median of (i) values to the right of the median, and (ii) values to the left of the median, then calculate the difference.

$$\text{IQR} = \left(\frac{4.03 + 4.10}{2}\right) - \left(\frac{3.65 + 3.74}{2}\right)$$

$$= 0.37 \text{ minutes}$$

Step 7: Calculate 'sum of squares of values', then substitute the values into the computational formula.

$$s = \sqrt{\frac{1}{11}\left(180.4006 - \frac{(46.44)^2}{12}\right)}$$

$$= 0.248 \text{ minutes (3 s.f.)}$$

iii The mean, as there are no apparent outliers.

The standard deviation, again as there are no apparent outliers and it is always quoted with the mean.

Example 1.9 A householder keeps a daily record of the number of letters received. The results for a sample of 100 days are summarised in the table.

Number of letters	0	1	2	3	4	5	6	7	8
Number of days	9	16	19	22	12	10	7	3	2

a Calculate values for the mode, range, median, interquartile range, mean and standard deviation.

b Identify the most appropriate measure of average and the most appropriate measure of spread. Justify your choices.

Step 1: Identify the value with the greatest frequency.

a Mode = 3 letters

Step 2: Identify the highest and lowest values, then calculate the difference.

Range $= 8 - 0 = 8$ letters

Step 3: Determine the cumulative frequency.

Number of letters (x)	0	1	2	3	4	5	6	7	8
Number of days (f)	9	16	19	22	12	10	7	3	2
Cumulative number of days (F)	9	25	44	66	78	88	95	98	100

Step 4: Calculate 'average of middle two values'.

$$\text{Median} = \frac{3 + 3}{2} = 3 \text{ letters}$$

Step 5: Identify the median of (i) values to the right of the median, and (ii) values to the left of the median, then calculate the difference.

$$\text{IQR} = \left(\frac{4 + 4}{2}\right) - \left(\frac{1 + 2}{2}\right) = 2.5 \text{ letters}$$

Tip:
When finding median and quartiles, imagine all the individual values written as an ordered list.

Step 6: Use a table to record values for $\sum fx$ and $\sum fx^2$.

Letters (x)	Days (f)	fx	fx^2
0	9	0	0
1	16	16	16
2	19	38	76
3	22	66	198
4	12	48	192
5	10	50	250
6	7	42	252
7	3	21	147
8	2	16	128
Total	100	297	1259

Recall:
In fx^2, only the x is squared.

Step 7: Substitute the values into the formulae.

$$\bar{x} = \frac{\sum fx}{\sum f} = \frac{297}{100} = 2.97 \text{ letters}$$

$$s = \sqrt{\frac{1}{\sum f - 1}\left(\sum fx^2 - \frac{(\sum fx)^2}{\sum f}\right)} = \sqrt{\frac{1}{99}\left(1259 - \frac{297^2}{100}\right)}$$

$$= 1.95 \text{ letters (3 s.f.)}$$

Tip:
Use your calculator to find $\bar{x}$ and s directly, taking care with data entry.

b As there are no obvious outliers, the mean should be used as the measure of average and the standard deviation as the measure of spread.

SKILLS CHECK 1C: Choice of numerical measures

1 The number of cars in a small private car park at 9 am is recorded over a 3-week period with the following results.

11 12 16 15 14 13 14 17 18 10 14 19 13 14 20

  a Calculate the mode, median and mean.
  b Calculate the range, interquartile range and standard deviation.
  c State, with a reason, for each of **a** and **b**, which, if any, is the most appropriate measure.

2 A regular viewer of a TV quiz show keeps a record, for a sample of 20 shows, of the number of questions she answers correctly. The numbers are recorded as follows.

8 10 6 9 7 10 5 7 8 36 12 14 8 10 8 15 14 33 13 7

  a Calculate the mode, median and mean.
  b Calculate the range, interquartile range and standard deviation.
  c State, with a reason, for each of **a** and **b**, which, if any, is the most appropriate measure.

3 The data below show the weights, in kilograms, of a sample of 10 bundles of firewood.

2.14 2.08 2.24 2.09 2.13 2.08 2.20 2.02 2.17 2.05

  a Calculate the median and mean.
  b Calculate the range, interquartile range and standard deviation.
  c State, with a reason, for each of **a** and **b**, which, if any, is the most appropriate measure.

4 The time taken, in hours, for a postman to complete his rural van round is shown for a sample of 12 days.

2.15 2.59 4.82 2.03 2.23 2.47 2.73 2.74 2.56 2.19 2.11 2.09

a Calculate the median and mean.

b Calculate the range, interquartile range and standard deviation.

c State, with a reason, for each of **a** and **b**, which, if any, is the most appropriate measure.

d In fact, the value of 4.82 is an incorrect recording of 2.84. Without further calculations:

   i state, with reasons, whether or not this change affects your answers to **a** and **b**;

   ii state, with a reason, whether or not this change affects your answers to **c**.

5 A count of the number of matches, in each of a random sample of 48 boxes, results in the following frequency distribution.

Number of matches	49	50	51	52	53
Number of boxes	2	31	8	5	2

a Calculate the mode, median and mean.

b Calculate the range, interquartile range and standard deviation.

c State, with a reason, for each of **a** and **b**, which, if any, is the most appropriate measure.

6 The number of washers in each of a random sample of 72 packets is counted with the following results.

Number of washers	18	19	20	21	22	23	24
Number of packets	3	8	39	11	7	3	1

a Calculate the mode, median and mean.

b Calculate the range, interquartile range and standard deviation.

c State, with a reason, for each of **a** and **b**, which, if any, is the most appropriate measure.

 7 The table summarises the results of an investigation into the number of thistles in each of 100 randomly selected 1-metre square plots of grazing land on a farm.

Number of thistles	0	1	2	3	4	5	6 to 9	10 to 14	15 to 19	20 to 30
Number of plots	23	19	15	12	9	6	6	5	3	2

a Estimate the median and mean.

b Estimate the interquartile range and standard deviation.

c State, with a reason, for each of **a** and **b**, which, if any, is the most appropriate measure.

SKILLS CHECK 1C EXTRA is on the CD

Examination practice   Numerical measures

1 A crate contains 20 trays of eggs, each tray holding 30 eggs. The total weight, in grams, of the eggs on each tray is shown below.

2147 2258 2456 2371 2078 2168 2247 2289 2215 2148
2251 2361 2113 2230 2351 2263 2176 2340 2468 2230

For this crate, calculate the mean and standard deviation for the total weight of eggs on a tray.

2 A library contains a large number of shelves all of the same length. A count is made of the number of books on each of a random sample of 25 shelves with the following results.

49 38 25 46 17 50 35 42 45 27 45 39 27 43 18 47 38 41 26 35 43 20 29 36 44

Estimate the mean and standard deviation of the number of books on a shelf in this library.

3 A garden fence consists of 23 panels each of length 2 metres and each containing 13 vertical boards. The number of faulty boards in each panel is counted with the following results.

0 2 0 2 4 3 0 0 1 1 0 1 0 0 2 1 0 1 0 2 1 0 2

a Calculate the mode, median and mean.

b Calculate the range, interquartile range and standard deviation.

c State, with a reason, for each of **a** and **b**, which, if any, is the most appropriate measure.

4 A firm is contemplating the introduction of an aptitude test, consisting of a puzzle, for future applicants for work on the assembly line. To assess the test, current employees on the assembly line were asked to complete the puzzle. The recorded times, x seconds, to the nearest second, of 35 such employees are shown below. These times have been ordered with the quickest first and slowest last.

26 29 31 32 35 38 40 41 43 44 45 46 48 48 50 50 51 52
53 53 54 55 56 56 58 59 62 66 71 76 83 85 92 94 97

a Calculate the median and interquartile range.

b Given that $\sum x = 1919$ and $\sum(x - \bar{x})^2 = 11591$, calculate the mean and standard deviation.

c In addition to the times above, six other current employees on the assembly line took more than 100 seconds, the time at which recordings ceased, to complete the puzzle. Calculate the median time for all 41 current employees on the assembly line.

d The firm decides not to offer employment to applicants who take longer than the average time taken by the employees who took the test.

    i State, with justification, whether you would recommend the median or the mean to be used as the measure of average.

    ii Write down the value of your recommended measure of average.

5 A random sample of 150 inner-city shop assistants was asked how many of their six most recent journeys into work had been made by bus. The results were as follows.

Number of journeys	0	1	2	3	4	5	6
Number of shop assistants	18	6	5	16	14	28	63

a Calculate the mode, median, range and interquartile range.

b State, with a reason, which of those in **a** are the most appropriate measures.

c Why are the mean and standard deviation not appropriate measures for the above data?

 6 The basic weekly earnings, to the nearest £, of 150 workers in a factory are shown in the following table.

Weekly earnings (£)	100–109	110–129	130–149	150–174	175–204
Number of workers	12	35	62	28	13

a Calculate estimates of the mean and standard deviation of this distribution.

b The union, negotiating a pay increase for the workers, puts forward two proposals to the management of the factory.

> **Proposal 1:** A 5% pay increase.
>
> **Proposal 2:** A 3% basic pay increase plus a fixed lump sum of £156 per worker **per annum**.

- **i** For **each** proposal, state new estimates of the mean and standard deviation of the basic weekly earnings.
- **ii** The management decide to accept one of the above proposals but wish to minimise their wage bill. Which of the proposals should the management accept? Justify your answer. [AQA(NEAB) June 1994]

Please note that this question is NOT from the live examinations for the current specification.

7 The numbers below represent the scores of batsmen during the 2004 cricket season.

> 15 26 0 4 7 48 73 0 12 15 47 0 190 17 0 5 13 21 115 1 2 73 24 5 0 88 17

a Calculate the mode, median and mean.

b Calculate the range, interquartile range and standard deviation.

c State, with a reason, for each of **a** and **b**, which, if any, is the most appropriate measure.

8 The ages, in years, of guests staying at a five-star hotel for a gourmet weekend are summarised in the table.

Age (years)	Number of guests
18–21	2
22–29	3
30–39	15
40–49	22
50–69	28
70–79	14
Total	**84**

a Calculate estimates for the mean and standard deviation of these data.

b The ages of guests staying at the hotel at other times have a mean of 39.3 years and a standard deviation of 17.8 years.

- **i** Compare, briefly, the ages of guests staying for the gourmet weekend with those of guests staying at other times.
- **ii** Suggest a reason for any differences.

 9 A student records the maximum daily temperature, y °C, in the garden of her home during July 2000 with the following results.

> 20 21 24 21 20 23 22 21 20 18 19 22 22 24 26 24
> 27 28 30 29 27 24 22 21 24 22 24 25 24 20 19

a Calculate the mean and standard deviation of these temperatures.

b From weather records, the student discovers that, during July 1900, the mean and standard deviation of the maximum daily temperature in the region including her home was 70.7 °F and 8.1 °F, respectively. Given that to convert x °F to y °C, the formula is $y = \dfrac{5}{9}x - \dfrac{160}{9}$, compare the maximum daily temperatures in the region for July 2000 with those for July 1900.

2 Probability

2.1 Introduction

Elementary probability; the concept of a random event and its probability.

A **statistical experiment** will often consist of a set of **trials**. Each trial will result in an outcome, or set of outcomes, called an **event**.

The **probability of event** A, denoted by P(A), is measured on a scale from 0 to 1. Zero represents impossibility and unity represents certainty.

When a trial results in N **equally likely outcomes**, of which $n(A)$ result in the event A, then

$$P(A) = \frac{n(A)}{N}$$

Example 2.1 A fair die is thrown. Find the probability that the score is:

a 3   **b** even   **c** more than 2.

Step 1: Find the total number of equally likely outcomes, and hence the probability of each outcome.

Step 2: Find the number of outcomes that result in the event.

Step 3: Calculate the probability of the event.

There are six equally likely outcomes so the probability of each outcome is $\frac{1}{6}$. Thus

a $P(\text{score} = 3) = \frac{1}{6}$

b $P(\text{score} = \text{even}) = P(\text{score} = 2, 4 \text{ or } 6) = \frac{3}{6} = \frac{1}{2}$

c $P(\text{score} > 2) = P(\text{score} = 3, 4, 5 \text{ or } 6) = \frac{4}{6} = \frac{2}{3}$

> **Note:**
> An event can consist of more than one outcome.

In situations where equally likely outcomes cannot be assumed, then P(A) may be estimated by the **relative frequency of event** A, where:

$$\text{Relative frequency } (A) = \frac{\text{Number of times event } A \text{ occurs}}{\text{Total number of trials}}$$

> **Note:**
> In examination questions, probabilities will either be given or be capable of determination using equally likely outcomes or relative frequencies.

Example 2.2 The following table shows the contents of a small bookcase.

	Statistics	General
Paperback	12	13
Hardback	18	7

> **Note:**
> Summarising information in such tables may sometimes be helpful in answering probability questions based on relative frequencies.

A book is selected at random from the bookcase. Find the probability that the book is:

a paperback   **b** general   **c** statistics hardback.

Step 1: Find the table's row and column totals

	Statistics	General	Total
Paperback	12	13	25
Hardback	18	7	25
Total	30	20	50

> **Note:**
> Always check that both row totals and column totals add to the same total.

a P(paperback) $= \frac{25}{50} = \frac{1}{2}$ or 0.5

b P(general) $= \frac{20}{50} = \frac{2}{5}$ or 0.4

c P(statistics hardback) $= \frac{18}{50} = \frac{9}{25}$ or 0.36

2.2 Addition law

Addition law of probability. Mutually exclusive events.

Given any event A, then the event 'A does not occur' is called its
complementary event and is denoted by A'. Since either event A
occurs or does not occur

$$P(A) + P(A') = 1 \quad \text{or} \quad P(A') = 1 - P(A)$$

Also, events A and A' are said to be **mutually exclusive**, since they
cannot occur together, and **exhaustive**, as one of them must occur.

If A and B are **two mutually exclusive events**, then

$$P(A \text{ or } B) \equiv P(A \cup B) = P(A) + P(B)$$

> **Note:**
> This is the **addition law for two mutually exclusive events**. It can be extended to any number of mutually exclusive events.

If A and B are **any two events**, then

$$P(\text{at least one of } A \text{ or } B) \equiv P(A \text{ or } B \text{ or both})$$
$$= P(A \cup B)$$
$$= P(A) + P(B) - P(A \cap B)$$

where

$$P(A \cap B) \equiv P(A \text{ and } B) \equiv P(\text{both } A \text{ and } B \text{ occur})$$

> **Note:**
> $A \cup B$ is the **union** of A and B.

> **Note:**
> This is the **addition law for any two events** and is provided in the examination. Extensions to more than two events will **not** be required in the examination.

> **Note:**
> $A \cap B$ is the **intersection** of A and B. For mutually exclusive events A and B, $P(A \cap B) = 0$.

Venn diagrams may be used to illustrate these results and may
sometimes be helpful in answering probability questions.

> **Note:**
> In the examination, you will be expected to understand set notation but its use will **not** be required.

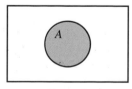

Event A

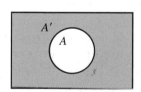

Event A'

> **Note:**
> $P(A' \cap B') = 1 - P(A \cup B)$.

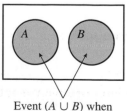

Event ($A \cup B$) when
mutually exclusive

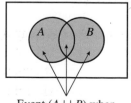

Event ($A \cup B$) when
not mutually exclusive

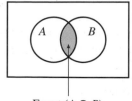

Event ($A \cap B$)

Example 2.3 A card is selected at random from a pack of 52 playing cards. Determine the probability that the card is:

    a not a jack,
    b a king or an ace,
    c a jack or a queen or a king,
    d a heart or a five.

Step 1: Find the numbers of possible outcomes.

Step 2: Divide by 52, the total number of cards in a pack.

a $P(\text{jack}) = \frac{4}{52} = \frac{1}{13}$ so $P(\text{not a jack}) = 1 - \frac{1}{13} = \frac{12}{13}$

b Events 'king or ace' are mutually exclusive, so

$$P(\text{king or ace}) = P(\text{king}) + P(\text{ace}) = \frac{4}{52} + \frac{4}{52} = \frac{8}{52} = \frac{2}{13}$$

c Events 'jack or queen or king' are mutually exclusive, so

$$P(\text{jack or queen or king}) = P(\text{jack}) + P(\text{queen}) + P(\text{king})$$
$$= \frac{4}{52} + \frac{4}{52} + \frac{4}{52} = \frac{12}{52} = \frac{3}{13}$$

d Events 'heart or five' are not mutually exclusive (five of hearts), so

$$P(\text{heart or five or both}) = P(\text{heart}) + P(\text{five}) - P(\text{five of hearts})$$
$$= \frac{13}{52} + \frac{4}{52} - \frac{1}{52} = \frac{16}{52} = \frac{4}{13}$$

> **Note:**
> When using the addition law, always first check whether or not the events are mutually exclusive.

SKILLS CHECK **2A: Addition law**

1 The mutually exclusive events A and B are such that $P(A) = 0.55$ and $P(B) = 0.35$.
    a State values for $P(A')$ and $P(B')$.
    b Find $P(A \cup B)$ and $P(A' \cap B')$.
    c State, giving a reason whether A and B are exhaustive events.

2 The events C and D are such that $P(C') = 0.30$, $P(D) = 0.55$ and $P(C \cap D) = 0.40$.
    a State values for $P(C)$ and $P(D')$.
    b Find $P(C \cup D)$.
    c Find $P(C' \cap D')$ and hence $P(C' \cup D')$.

3 The events A, B, C and D are mutually exclusive and exhaustive events with $P(A) = 0.25$, $P(B) = 0.20$ and $P(C) = 0.40$.
    a Find $P(D)$.
    b Find $P(A \cup C \cup D)$.
    c Are the events A' and B' mutually exclusive? Justify your answer.

4 A new housing development contains 100 properties of which 40 are semi-detached houses, 25 are detached houses, 20 are semi-detached bungalows and 15 are detached bungalows. A property is selected at random. Determine the probability that the property selected is:
    a a house;
    b detached;
    c either detached or a bungalow or both.

5 A card is selected at random from a pack of 52 playing cards with aces scored as ones. Calculate the exact probability that the card is:
    a a heart;
    b not a club;
    c a picture;
    d either a heart or a picture or both;
    e either a heart or a picture but not both;
    f less than 5;
    g red or above 8 or both.

6 A box of mixed screws contains:

4 screws of length 25 mm;
6 screws of length 30 mm;
10 screws of length 35 mm;
12 screws of length 40 mm;
10 screws of length 45 mm;
8 screws of length 50 mm.

A screw is selected at random from the box. Calculate the probability that the length of the screw is:

a exactly 40 mm;   **b** more than 30 mm;

c at most 40 mm;   **d** at least 40 mm;

e at least 30 mm but at most 40 mm;   **f** not 35 mm;

g not less than 45 mm.

7 An unbiased 20-sided die, with faces marked with the numbers 1 to 20, is rolled.

A denotes the event 'score is odd'.
B denotes the event 'score is divisible by 5'.
C denotes the event 'score is less than 10'.
D denotes the event 'score is 11 or more'.

a Calculate:

  i $P(B \cup C)$ **ii** $P(A \cup D)$.

  iii $P(A \cap C)$ **iv** $P(B \cap D)$

b Identify:

  i two events that are mutually exclusive;

  ii three events that are exhaustive.

8 The population of a housing estate, classified by gender and age, is shown in the table.

	Under 18	18 to 64	65 or over
Male	39	63	18
Female	27	71	32

A person is selected at random from those living on the estate.

M denotes the event 'person selected is male'.
C denotes the event 'person selected is under 18'.
R denotes the event 'person selected is 65 or over'.

a Calculate:

  i $P(R)$ **ii** $P(M')$ **iii** $P(M \cup C)$

  iv $P(M \cup R')$ **v** $P(C \cup R)'$

b Describe, concisely in context, the event:

  i $M \cup R'$ **ii** $M' \cap (C \cup R)'$

c For events *M*, *C* and *R*, write down:

  i two that are mutually exclusive;

  ii two that are not mutually exclusive.

SKILLS CHECK **2A EXTRA** is on the CD

2.3 Multiplication law

Multiplication law of probability and conditional probability. Independent events.

Events A and B are said to be **independent** when the probability of event A occurring is not affected by whether or not event B occurs.

If A and B are **two independent events**, then

$$\text{P}(A \text{ and } B) \equiv \text{P}(A \cap B) = \text{P}(A) \times \text{P}(B)$$

If A and B are **any two events**, then

$$\text{P}(A \text{ and } B) \equiv \text{P}(A \cap B) = \text{P}(A) \times \text{P}(B|A) = \text{P}(B) \times \text{P}(A|B)$$

where $\text{P}(B|A)$ denotes the **conditional probability of B given A.**

> **Note:**
> This is the **multiplication law for two independent events**. It can be extended to any number of independent events.

> **Note:**
> This is the **multiplication law for any two events** and is provided in the examination. It can be extended to any number of events.

> **Note:**
> For independent events A and B, $\text{P}(B|A) = \text{P}(B)$.

Tree diagrams may be used to illustrate these results and may be helpful in answering probability questions.

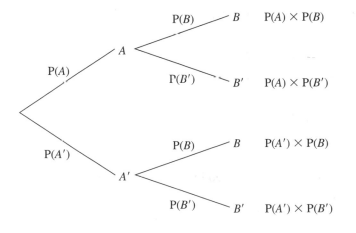

Independent events

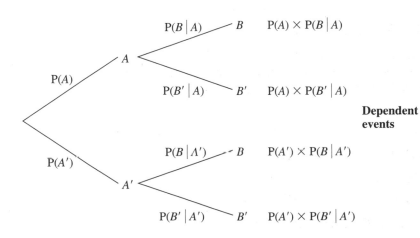

Dependent events

> **Note:**
> In any complete tree diagram, the sum of the probabilities at the end of the tree branches should be 1.

> **Note:**
> When using the multiplication law, always first check whether or not the events are independent.

Example 2.4 A bag contains 2 red (R) discs and 3 blue (B) discs. Two discs are drawn at random, with replacement, from the bag. Calculate the probability that:

a the first disc is red and the second disc is blue;

b both discs are blue;

c the two discs are the same colour.

> **Note:**
> The term 'with replacement' implies that the first disc is returned to the bag before the second disc is drawn. As a result the drawings are **independent**.

Step 1: Use the multiplication law for two independent events.

a $P(R \text{ and } B) = \frac{2}{5} \times \frac{3}{5} = \frac{6}{25}$ or 0.24

b $P(B \text{ and } B) = \frac{3}{5} \times \frac{3}{5} = \frac{9}{25}$ or 0.36

Step 2: Use the addition law for two mutually exclusive events.

c $P(\text{same colour}) = P((R \text{ and } R) \text{ or } (B \text{ and } B))$

$= P(R \text{ and } R) + P(B \text{ and } B)$

$= (\frac{2}{5} \times \frac{2}{5}) + \frac{9}{25} = \frac{13}{25}$ or 0.52

Recall:
The events $(R \text{ and } R)$ and $(B \text{ and } B)$ are mutually exclusive, as they cannot occur together.

Example 2.5 A bag contains 4 red (R) balls, 5 white (W) balls and 1 blue (B) ball. Two balls are selected at random, without replacement, from the bag. Calculate the probability that:

a the first ball is red and the second ball is white;

b both balls are blue;

c exactly one of the two balls is red;

d the two balls are different colours.

Note:
The term 'without replacement' implies that selected balls are not returned to the bag before the next ball is selected. As a result the drawings are **dependent**.

Step 1: Draw the branches for the tree diagram.

Step 2: Enter the probabilities on the branches.

Step 3: Multiply out these probabilities for each route through the diagram.

Step 4: Check that the final probabilities total 1.

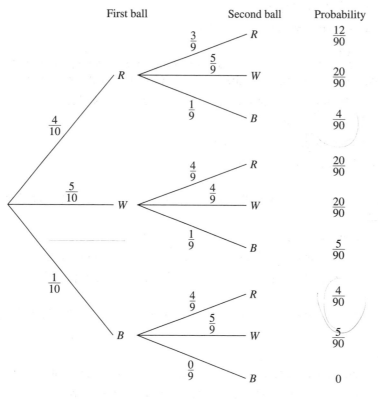

Note:
After the first ball is selected, there are only nine balls remaining in the bag.

Note:
As there is only one blue ball, the event $(B \cap B)$ is impossible.

Step 5: Use the tree diagram.

a $P(R \cap W) = \frac{20}{90} = \frac{2}{9}$ or 0.222 (3 s.f.)

Recall:
This is equivalent to $P(R) \times P(W \mid R)$.

Step 6: Use the addition law for two mutually exclusive events and the tree diagram.

b $P(B \cap B) = 0$

Recall:
The probability of an impossible event is zero.

c $P(1 \text{ red}) = P(R \cap R') + P(R' \cap R) = (\frac{20}{90} + \frac{4}{90}) + (\frac{20}{90} + \frac{4}{90})$

$= \frac{48}{90} = \frac{24}{45}$ or 0.533 (3 s.f.)

Note:
$P(A \cap B) = P(B \cap A)$.

Step 7: For the complementary event, use the addition law for two mutually exclusive events and the tree diagram.

d $P(\text{different colours}) = 1 - P(\text{same colour})$

$= 1 - [P(R \cap R) + P(W \cap W) + P(B \cap B)]$

$= 1 - (\frac{12}{90} + \frac{20}{90} + 0)$

$= 1 - \frac{32}{90} = \frac{58}{90} = \frac{29}{45}$ or 0.644 (3 s.f.)

Recall:
$P(A) = 1 - P(A')$.

1 The independent events A, B and C are such that $P(A) = 0.55$, $P(B) = 0.40$ and $P(C) = 0.20$.

    a Find $P(A \cap B)$ and $P(A \cap B \cap C)$.

    b Find $P(B' \cap C)$ and $P(A' \cap B' \cap C')$.

2 The events E and F are such that $P(E) = 0.60$, $P(F) = 0.30$ and $P(E \mid F) = 0.20$.

    a Determine $P(E \cap F)$.

    b Determine $P(F \mid E)$.

3 A bag contains 2 red balls, 3 white balls and 5 blue balls. Three balls are selected at random, with replacement, from the bag. Calculate the probability that:

    a the first ball selected is red, the second is white and the third is blue;

    b the balls selected are different colours;

    c all three balls selected are red;

    d all three balls selected are the same colour.

 4 A box contains 4 blue pens, 3 black pens, 2 red pens and 1 green pen. Two pens are randomly selected, without replacement, from the box. Calculate the probability that:

    a the first pen selected is blue and the second pen selected is black;

    b both pens selected are black;

    c both pens selected are the same colour;

    d neither pen selected is blue or black.

5 Two cards are randomly chosen, with replacement, from a pack of 52 playing cards. Calculate the probability that:

    a both cards chosen are red;

    b the first card chosen is red and the second card chosen is a picture;

    c one card chosen is a club and the other card chosen is a multiple of 4.

6 Two cards are randomly chosen, without replacement, from a pack of 52 playing cards. Calculate the probability that:

    a both cards chosen are red;

    b the first card chosen is a multiple of 3 and the second card chosen is a multiple of 5;

    c one card chosen is an ace and the other card chosen is a picture.

 7 Vehicles approaching a roundabout from the north must go in one of three directions: east (left), south (straight on) or west (right). A traffic survey establishes that 65% of vehicles approaching the roundabout from the north turn left, 15% go straight on and 20% turn right. It may be assumed that drivers of vehicles approaching from the north make independent decisions on their direction of travel at the roundabout. Determine the probability that, for the next three vehicles approaching the roundabout from the north:

    a all turn left;

    b all go in the same direction;

    c the first turns left, the second goes straight on and the third turns right;

    d all go in different directions;

    e exactly two turn left;

    f none go straight on.

8 A manufacturer of fridges uses condensers from three suppliers, **K**, **L** and **M**, in the ratios $4:5:1$, respectively. Of the condensers from supplier **K**, 8 per cent fail within the warranty period. The corresponding figures for the condensers supplied by suppliers **L** and **M** are 4 per cent and 10 per cent, respectively. For a fridge selected at random, calculate the probability that its condenser:

a was supplied by **L** and did not fail during the warranty period;

b was supplied by **M** and failed during the warranty period;

c failed during the warranty period;

d was supplied by **K**, given that it failed during the warranty period.

SKILLS CHECK **2B EXTRA** is on the CD

2.4 Application of probability laws

Application of probability laws.

Example 2.6 The table shows the numbers of males and females in each of four employment categories in an engineering company.

	Employment category			
	White collar	**Blue collar**	**Skilled**	**Manual**
Male	13	33	139	25
Female	2	7	26	5

An employee is selected at random.

F denotes the event 'the employee is female'.
W denotes the event 'employee is white collar'.
B denotes the event 'employee is blue collar'.
S denotes the event 'employee is skilled'.

a Determine:

i $P(F)$ **ii** $P(F \cap B)$ **iii** $P(F' \cup S)$ **iv** $P(F'|B)$

b Four employees are randomly selected without replacement. Determine the probability that exactly one employee is female.

c Describe in context, as simply as possible, the events denoted by:

i $F' \cap S$ **ii** $F' \cup B$

> **Note:**
> Answers can often be found directly from a table without the need to use the two probability laws.

Step 1: Add the row entries and column entries and insert suitable notation.

	Employment category				
	White collar (W)	**Blue collar** (B)	**Skilled** (S)	**Manual**	**Total**
Male	13	33	139	25	210
Female (F)	2	7	26	5	40
Total	15	40	165	30	250

Step 2: Use the table to find the required probabilities.

a i $P(F) = \frac{40}{250} = \frac{4}{25}$ or 0.16 **ii** $P(F \cap B) = \frac{7}{250}$ or 0.028

iii $P(F' \cup S) = \frac{210 + 26}{250} = \frac{236}{250} = \frac{118}{125}$ or 0.944

iv $P(F' \mid B) = \frac{33}{40}$ or 0.175

> **Recall:**
> F' denotes the event 'not F'.

Step 3: Find the probability of *F F′ F′ F′* then multiply by 4 to allow for the different positions that *F* can occur.

b $P(1 \text{ female in sample of } 4) = \frac{40}{250} \times \frac{210}{249} \times \frac{209}{248} \times \frac{208}{247} \times 4$
$$= 0.383 \text{ (3 s.f.)}$$

Step 4: Translate the set notation into context.

c i Employee is a skilled male.

  ii Employee is either male or blue collar (or both).

Example 2.7 A popular car is available in a variety of models with 30% of them being three-door hatchbacks, 55% of them being five-door hatchbacks and the remainder being coupés.

Of the three-door hatchbacks, 60% are fitted with a 1.1-litre petrol engine, 15% are fitted with a 1.4-litre petrol engine and the remaining 25% are fitted with a diesel engine.

The corresponding figures for the five-door hatchbacks are 15%, 55% and 30%, respectively, and for coupés are 5%, 85% and 10%, respectively.

a One of these popular cars is chosen at random. Find the probability that:

  i it is a five-door hatchback fitted with a 1.4-litre petrol engine;

  ii it has a diesel engine;

  iii it is a coupé or it is fitted with a 1.4-litre petrol engine;

  iv it is a coupé, given that it is fitted with a 1.4-litre petrol engine.

b Two of these popular cars are chosen at random. Find the probability that they have the same type of engine.

Step 1: Draw the branches for the tree diagram.

Step 2: Enter the probabilities on the branches.

Step 3: Multiply out these probabilities for each route through the diagram.

Step 4: Check that the final probabilities total 1.

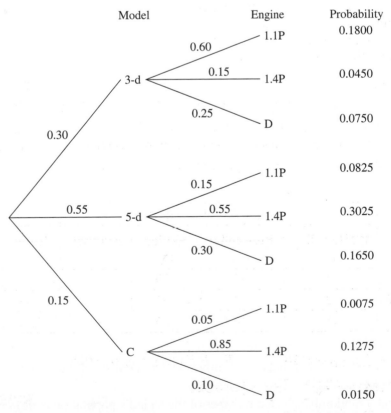

Note:
In questions such as this, a tree diagram makes subsequent calculations much easier.

Recall:
In any complete tree diagram, the sum of the probabilities at the end of the tree branches should be 1.

Total = 1

Step 5: Use the tree diagram.

a i $P(5\text{-}d \cap 1.4P) = 0.3025$

ii $P(D) = 0.0750 + 0.1650 + 0.0150 = 0.255$

iii $P(C \cup 1.4P) = 0.0450 + 0.3025 + 0.15 = 0.4975$

Step 6: Also use

$$P(B \mid A) = \frac{P(A \cap B)}{P(A)}.$$

iv $P(C \mid 1.4P) = \dfrac{P(C \cap 1.4P)}{P(1.4P)} = \dfrac{0.1275}{0.0450 + 0.3025 + 0.1275}$

$$= \frac{0.1275}{0.475} = 0.268 \text{ (3 s.f.)}$$

> **Note:**
> Rearrangement of multiplication law.

Step 7: Use the multiplication law for two independent events on each of three mutually exclusive events.

b $P(2 \text{ cars have same type of engine}) = [P(1.1P)]^2 + [P(1.4P)]^2 + [P(D)]^2$

$$= 0.27^2 + 0.475^2 + 0.255^2$$

$$= 0.364 \text{ (3 s.f.)}$$

> **Note:**
> $0.27 + 0.475 + 0.255 = 1$, as it must.

SKILLS CHECK **2C: Application of probability laws**

1 At the start of a game of darts, Les and Kirk independently throw alternate darts until one of them hits a 'double'. The probability that Les hits a 'double' with any dart is 0.4, and the probability that Kirk hits a 'double' with any dart is 0.2.

   a Given that Les throws the first dart, show that the probability that Kirk misses a 'double' with his first throw is 0.48.

   b Find the probability that Les hits a 'double' on:

     i his third throw   **ii** his fifth throw.

2 Over a period of time, a commuter establishes that the probability that her train is cancelled is 0.03. When her train is not cancelled, the probability that it is on time is 0.85.

   a Determine the probability that her train is on time.

   b Given that her train is not on time, determine the probability that it has been cancelled.

 3 Two letters, **F** and **S**, are posted in the same post box on the same day to the same address. Letter **F** is sent by first-class post and has a probability of 0.95 of being delivered the next day. Letter **S** is sent by second-class post and has a probability of only 0.15 of being delivered the next day.

   a Find the probability that:

     i neither letter is delivered the next day;

     ii at least one of the letters is delivered the next day.

   b Given that at least one of the letters is delivered the next day, calculate the probability that letter **F** is delivered the next day.

4 A workshop contains 10 machines: 5 of type **R**, 3 of type **S** and 2 of type **T**. All machines, irrespective of type, are equally reliable. Given that **three different** machines break down during the same week, calculate the probability that:

   a all arc type **R**;   **b** all are the same type;   **c** none is type **S**;

   d the first to break down is type **R**, the second is type **S** and the third is type **T**;

   e no two are the same type.

5 It is known that 65% of an island's population have been inoculated against a particular virus. Of those who have been inoculated, 95% are then immune to the virus. Of those who have not been inoculated, only 15% are immune to the virus.

   a Calculate the probability that a randomly selected member of the island's population is immune to the virus.

   b Given that a randomly selected member of the island's population is immune to the virus, find the probability that the member has been inoculated.

6 A university cricket club has 50 players, of whom 36 are British. The table shows the type of players at the club.

	Batsman	All-rounder	Wicket keeper	Bowler
British	16	4	2	14
Non-British	6	2	1	5

A player is selected at random.

a Find the probability that the player selected is:

i British;

ii an all-rounder;

iii a British bowler;

iv an all-rounder or a wicket-keeper;

v British or a batsman or both;

vi British or a batsman but not both.

b i Given that the selected player is British, find the probability that he is a bowler.

ii Given that the selected player is a bowler, find the probability that he is British.

7 The publishers of a journal conducted a survey of a sample of its subscribers, exactly 70% of the sample being male. In the survey, the subscribers were asked to place themselves in the most appropriate of four categories: student, employed, unemployed, retired. Of the male subscribers in the sample, 10% were students, 65% were employed, 10% were unemployed and 15% were retired. For the female subscribers in the sample, the corresponding percentages were 5%, 45%, 35% and 15%.

a Determine the probability that a subscriber, chosen at random from those in the sample:

i is an employed male;

ii is retired;

iii is female or is retired or both;

iv is female, given that the subscriber is retired.

b Assuming that the sample selected for the survey is representative of all the journal's subscribers, calculate the probability that a random sample of 4 subscribers contains exactly 2 who are employed.

 8 An appliance repair company constructed the following two-way frequency table from an examination of its records on the repair of washing machines.

	Cold water fill only	Hot and cold water fill
Without dryer	60	105
With dryer	20	95

a Calculate the probability that a washing machine repair record, chosen at random, is for a machine that:

i has hot and cold water fill but has no dryer;

ii has hot and cold water fill or a dryer or both;

iii has hot and cold water fill, given that it has a dryer.

b Further examination of the above repair records show that the number of washing machines repaired under guarantee comprised:

15% of the machines with cold water fill only and with no dryer;

20% of the machines with hot and cold water fill but with no dryer;

40% of the machines with cold water fill only but with a dryer;

60% of the machines with hot and cold water fill and with a dryer.

Calculate the probability that a washing machine repair record, chosen at random, is for a machine that:

i was repaired under guarantee;

ii has hot and cold water fill and a dryer, given that it was repaired under guarantee.

9 A professional society has four categories of membership: chartered, graduate, fellow and student. The number of members in each category, classified by gender, is shown in the table:

	Chartered	Graduate	Fellow	Student
Male	564	263	109	64
Female	282	81	81	56

One member is selected at random to present a gift to an honoured guest at the annual dinner.

a Calculate the probability that:

i a female member is selected;

ii a student member is selected;

iii a male graduate member is selected;

iv a chartered member is selected, given that a female is selected;

v a female is selected, given that a chartered member is selected.

b The event 'a male member is selected' is denoted by M.
The event 'a chartered member is selected' is denoted by C.
The event 'a student member is selected' is denoted by S.
For these events:

i write down two that are mutually exclusive;

ii find, justifying your answer, two that are neither mutually exclusive nor independent.

iii find, justifying your answer, two that are independent.

SKILLS CHECK **2C EXTRA** is on the CD

Examination practice   Probability

1 In a group of students 60% are female and 40% are male. A third of the female students study Spanish but only a quarter of the male students study Spanish.

A student is chosen at random from the group.

a i Show that the probability that the chosen student is female and studies Spanish is 0.2.

ii Calculate the probability that the chosen student studies Spanish.

b Given that the chosen student **does** study Spanish, calculate the conditional probability that the chosen student is female.   [AQA(A) Nov 2002]

2 A customer goes into a store to buy a refrigerator and a microwave. From past experience it is known that 10% of the refrigerators and 5% of the microwaves will be found to be defective when tested. The customer chooses one refrigerator and one microwave at random and the items are tested.

a Find the probability that:

i both items are found to be defective;

ii neither item is found to be defective;

iii exactly one of the items is found to be defective.

b Given that exactly one of the items is found to be defective, find the probability that it is the refrigerator.   [AQA(A) Jan 2004]

3 A football club employs 30 players, of whom 13 have British nationality. The following table shows how many of each type of player are employed by the club.

	Goalkeepers	Defenders	Midfielders	Attackers
British	2	4	5	2
Other	2	5	7	3

One player is chosen at random from the 30 players.

a Find the probability that the chosen player is:

   i British;

   ii a goalkeeper;

   iii British and a goalkeeper;

   iv a defender or a midfielder;

   v British or an attacker but **not** both.

b i Given that the chosen player is British, find the probability that he is a goalkeeper.

   ii State, giving a reason, whether the events

      'the chosen player is British' and

      'the chosen player is a goalkeeper'

   are independent.
                                 [AQA(A) May 2004]

4 A tennis player is allowed two serves. The probability that a particular player's first serve will be in is 0.7. She has probability 0.8 of winning the point when her first serve is in. Otherwise, her probability of winning the point is only 0.4. This information is shown in the tree diagram.

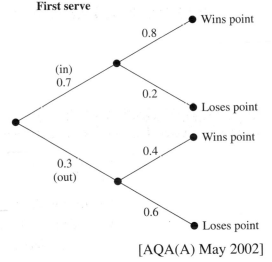

First serve

   a Find the probability that she wins a point when she is serving.

   b Given that she does win this point, find the probability that her first serve was in.

[AQA(A) May 2002]

5 A boy receives a school report card each week.
He is given a special treat whenever his report indicates 'Good behaviour' **and** 'Excellent homework'.
His behaviour is good with probability 0.6.
When his behaviour is good, he has a probability of 0.8 of doing excellent homework.
When his behaviour is **not** good, his probability of doing excellent homework is only 0.5.

   a For a random week, calculate the probability that:

     i he will be given a special treat;

     ii his homework will be excellent but he will **not** be given a special treat;

     iii his homework will be excellent.

   b Given that his homework is excellent, calculate the conditional probability that he is **not** given a special treat.
                                     [AQA(A) Jan 2003]

6 Jane is an athlete in training for a major event.

Every evening she does either a 5-mile run or a 10-mile run.

On a given day Jane does a full day's work with probability $\frac{2}{3}$.

When she has done a full day's work, the probability that she does a 10-mile run is $\frac{2}{5}$.

When she has **not** done a full day's work, the probability that she does a 10-mile run is $\frac{4}{5}$.

a For a random day, calculate the probability that:

  i she does a full day's work and a 10-mile run;

  ii she does **not** do a full day's work but she does a 10-mile run;

  iii she does a 10-mile run.

b Given that she does a 10-mile run, find the conditional probability that she has done a full day's work. [AQA(A) Nov 2004]

7 At a university, 60% of students are studying Arts subjects and the rest are studying Science subjects. Of the Arts students, 55% are female and of the Science students, 35% are female. Find the probability that a student, selected at random:

a is female and studying an Arts subject;

b is male and studying a Science subject;

c is male;

d is studying Science, given that the student is male. [AQA(B) Jan 2003]

 8 One hundred and fifty students at a large catering college have to choose between three tasks as part of their final assessment. The tasks involve cake baking, pastry skills or sweet making. A summary of their choices is given below.

	Male	Female
Cake baking	54	26
Pastry skills	27	18
Sweet making	9	16

A student is selected at random from those at the catering college.

C denotes the event that the selected student chooses cake baking.

D denotes the event that the selected student chooses pastry skills.

M denotes the event that the selected student is male.

(M' denotes the event 'not M'.)

a Find the value of

  i $P(M)$,   **ii** $P(M \cap C)$,   **iii** $P(M' \cup D)$,   **iv** $P(M \mid D)$.

b State whether the events M and D are independent, giving a reason for your answer.

It is known that 30 per cent of the male students and 40 per cent of the female students are over 21 years of age.

c A student is selected at random and found to be over 21 years of age. Find the probability that the student is male. [AQA(B) June 2001]

9 Maurice works at home. At 2 p.m. he decides to take a break to buy a copy of the Chronicle newspaper. There are three nearby newsagents: Arif, Bob and Carol. However by 2 p.m. they may have sold all their Chronicles and so have none available. The independent probabilities that they have a Chronicle available at 2 p.m. are:

Arif 0.4   Bob 0.7   Carol 0.25

a State the probability that Bob does not have a Chronicle available at 2 p.m.

b Find the probability that none of the three newsagents has a Chronicle available at 2 p.m.

c i Find the probability that Bob does not have a Chronicle available at 2 p.m. but Arif does.

  ii Find the probability that Carol does not have a Chronicle available at 2 p.m. but Arif does.

d Maurice decides to visit the newsagents in turn until he obtains a Chronicle or until he has visited all three. He tosses a coin. If it lands heads he will visit the three newsagents in the order Bob, Arif, Carol. If it lands tails he will visit them in the order Carol, Arif, Bob.

Find the probability that he will obtain a Chronicle from Arif.

[AQA(B) Jan 2004]

10 A newspaper published an article concerning a proposed European law which would force motorists to pay compensation and damages in all accidents with cyclists, regardless of who is to blame. The newspaper received 150 letters in response to the article. The following table shows the number of letters received, classified by the writers' usual forms of transport and their attitudes to the proposal.

| | | Attitude to the proposal | | |
		For	Against	Neutral
Usual	Car	3	67	9
form of	Bicycle	27	5	10
transport	Other	8	12	9

a The editor selects a letter at random to consider for publication. Find the probability that the letter selected is:

  i for the proposal;

  ii for the proposal, given that the writer usually travels by car.

b The editor selects three letters at random, without replacement. Find the probability that one letter is for the proposal, one against and one neutral.

c The proportions of letters suitable for publication were:

> 0.50 of those which were for the proposal;
> 0.25 of those which were against the proposal;
> 0.75 of those which were neutral.

  i Find the probability that a randomly selected letter is for the proposal and suitable for publication.

  ii Find the probability that a randomly selected letter is suitable for publication.

  iii Find the probability that at least three letters will have to be selected (at random, without replacement) before a letter is found which is suitable for publication from a writer who is for the proposal.

[AQA(B) Jan 2004]

3 Binomial distribution

3.1 Discrete random variables

Discrete random variables.

In statistics, a **variable** is a characteristic that is counted or measured.

When such a variable may be subject to (random) variation, it is called a **random** variable.

A random variable is said to be **discrete** when a list of (all) its possible values may be constructed. Here, and in fact in most cases, discrete random variables arise from **counting** a characteristic.

Recall:
Discrete data also arises from counting.

Random variables are denoted by italic upper case letters such as X, Y, Z. Values of a random variable are denoted by italic lower case letters such as x, y, z.

Thus, for example,

$$P(X = x) = 0.3$$

is read as

> 'The probability that the (discrete) random variable X takes the value x is 0.3'

Note:
This is sometimes shortened to $p(x) = 0.3$.

The (probability) **distribution** for a discrete random variable, X, is defined by the possible values, x, of X together with their corresponding probabilities, $p(x)$. These pairs of values may be displayed in a (probability) table or by a (probability) function.

Note:
This is analogous to a frequency distribution for discrete data.

When $P(X \leq x)$, rather than $P(X = x)$, is given for each value x of X, the result is called a (**cumulative**) **distribution function**.

Recall:
Cumulative frequency.

Probabilities for a discrete random variable may be found from its table or function (or from its cumulative distribution function.)

Example 3.1 The discrete random variable, X, has the following distribution.

x	0	1	2	3	4
$p(x)$	0.1	0.3	0.3	0.2	0.1

Note:
The probabilities $p(x)$ are always mutually exclusive and exhaustive.

Find the probability that a value of X:

a is more than 1

b is at most 3

c is at least 1 but is less than 4.

Tip:
You must be able to translate such requests using the signs $<$, $>$, $\leq$ and $\geq$.

Step 1: Find the required values of X.

Step 2: Add (or subtract) the corresponding probabilities.

a $P(X > 1) = p(2) + p(3) + p(4) = 0.6$

b $P(X \leq 3) = 1 - p(4) = 0.9$

c $P(1 \leq X < 4) = p(1) + p(2) + p(3) = 0.8$

Tip:
It is sometimes easier to find the probability of the complementary event and then subtract from 1.

Example 3.2 The distribution of the random variable, Y, is given by

$$P(Y = y) = \begin{cases} \dfrac{4 - y}{10} & y = 0, 1, 2, 3 \\ \\ 0 & \text{otherwise} \end{cases}$$

Note:

The second line in such expressions is often omitted.

Determine the probability that a value of Y:

a is 2   **b** is at least 2   **c** is either 1 or 2.

Note:

Questions similar to those in Examples 3.1 and 3.2 will **not** appear in the examination.

Step 1: Find the required values of Y.
Step 2: Substitute the values into the formula to find the probabilities.
Step 3: Add (or subtract) these probabilities as necessary.

a $P(Y = 2) = \dfrac{4 - 2}{10} = 0.2$

b $P(Y \geqslant 2) = 0.2 + \dfrac{4 - 3}{10} = 0.3$

c $P(Y = 1) + P(Y = 2) = \dfrac{4 - 1}{10} + 0.2 = 0.5$

3.2 Binomial random variables

Conditions for application of a binomial distribution.

The prefix 'bi' generally infers that two of something is involved. For example, bicycle (two wheels), biannual (twice annually), bilingual (two languages), biathlon (two events). In 'binomial' it indicates two outcomes.

A binomial random variable must satisfy the following conditions:
- there is a fixed number, n, of trials
- each trial results in one of two outcomes – success or failure
- the probability of a success, p, is constant from trial to trial
- the trials are independent.

Note:

In practice, the outcome with the smaller probability is usually called a 'success'.

The letters n and p denote the **parameters** of the distribution and

$$X \sim B(n, p)$$

is read as

'The (discrete) random variable, X, has a binomial distribution with parameters n and p.'

Note:

In statistics, the symbol '$\sim$' is usually shorthand for 'is distributed as'.

Example 3.3 For each of the random variables described below, state whether or not a binomial distribution is a suitable model. If it is, state values for n and p. If it is not, give a reason why.

a The number of heads observed when a fair coin is tossed 24 times.

b The number of throws of a fair die required to obtain 12 'sixes'.

c The number of milk chocolates in a sample of 5 chocolates selected, without replacement, from a box containing 20 milk chocolates and 15 plain chocolates.

Step 1: Check if binomial conditions are satisfied.

Step 2: If yes, quote values for n and p. If not, give a reason.

a Yes; $n = 24$, $p = 0.5$.

b No; n is not fixed.

c No; p is not constant.

1 For each of the random variables described below, state whether or not a binomial distribution is a suitable model. If it is, state values for n and p. If it is not, give a reason why.

   a The number of tosses of a fair coin needed to obtain 10 heads.

   b The number 'threes' obtained when a fair die is rolled 60 times.

2 For each of the random variables described below, state whether or not a binomial distribution is a suitable model. If it is, state values for n and p. If it is not, give a reason why.

   a The number of green jelly babies in a random sample of 10 jelly babies, selected with replacement, from a box containing 80 jelly babies of which 8 are green.

   b The number of red wine gums in a random sample of 10 wine gums, selected without replacement, from a box containing 72 wine gums of which 12 are red.

3 A utility company discovers that 65% of its customers pay their accounts by direct debit. Give one reason why B(20, 0.65) may not be a suitable model for 20 customers living in the same street.

 4 A large bin contains 6400 used golf balls of which 1120 are unusable. The random variable X denotes the number of unusable golf balls in a random sample of 10 balls selected, without replacement, from the bin. Explain why X may be approximated by a binomial random variable and give values for its parameters.

5 A display box contains 240 coloured erasers of which 96 are red.

   a A random sample of 24 erasers is selected, with replacement, from the box and a count made of the number, R, of red erasers obtained. Explain why R may be modelled by a binomial distribution, and give values for its parameters.

   b Erasers are selected at random, without replacement, from the box until six red erasers are obtained. Give two reasons why the number of erasers selected cannot be modelled by a binomial distribution.

 6 Wei, a trainee, starts work in a hospital's call centre for admissions. During her first week, the probability that she requires assistance with each telephone call is 0.35.

   a Given that Wei answers 60 telephone calls during her first week, suggest a probability model for the number of calls Wei is able to answer without assistance, stating any assumptions that you have made.

   b The random variable, W, denotes the number of telephone calls Wei answers until she has answered 20 without assistance. State, giving a reason, whether a binomial distribution is an appropriate model for W.

   c A random sample of 60 telephone calls, answered by Wei during her first 3 months, is monitored. Give a reason why a binomial distribution is unlikely to be an appropriate model for the number of calls Wei answered without assistance in this sample.

Calculation of probabilities using formula.

If $X \sim B(n, p)$ then

$$P(X = x) = \binom{n}{x} p^x (1 - p)^{n - x} \quad \text{for} \quad x = 0, 1, 2, \ldots, n$$

where $\binom{n}{x} = \dfrac{n!}{x!(n - x)!}$

Note:
This expression is provided in the examination.

Note:
Alternative notations for $\binom{n}{x}$ are ^nC_x or $_nC_x$ and, for a given n, its values form the corresponding row in Pascal's triangle.

Example 3.4 A fair six-sided die is rolled 12 times. Find the probability of:

    a exactly 2 sixes

    b at most 2 sixes

    c at least 1 six

    d between 5 and 7, inclusive, even scores.

Recall:
$k! = k \times (k - 1) \times (k - 2) \times \ldots \times 3 \times 2 \times 1$ and $0! = 1$.

Step 1: Identify the values for n and p for each of X and Y.

Let X = number of sixes and Y = number of even scores.

Then $X \sim B(12, \frac{1}{6})$ and $Y \sim B(12, \frac{1}{2})$.

Note:
This form of cancellation in $\binom{n}{x}$ is always possible and eases calculations.

Step 2: Identify the value or values of X required.

Step 3: Substitute the identified values into the formula.

a $P(X = 2) = \binom{12}{2}\left(\dfrac{1}{6}\right)^2\left(1 - \dfrac{1}{6}\right)^{12 - 2} = \dfrac{12!}{2!\,10!}\left(\dfrac{1}{6}\right)^2\left(\dfrac{5}{6}\right)^{10}$

$\qquad\qquad\quad = \dfrac{(12 \times 11) \times 10!}{(2 \times 1) \times 10!}\left(\dfrac{1}{6}\right)^2\left(\dfrac{5}{6}\right)^{10}$

$\qquad\qquad\quad = 66 \times 0.02778 \times 0.16151 = 0.296 \text{ (3 s.f.)}$

Tip:
Get to know how to use the $\binom{n}{x}$ function on your calculator.

b $P(X \leqslant 2) = P(X = 0) + P(X = 1) + P(X = 2)$

$\qquad\qquad\quad = \left(\dfrac{5}{6}\right)^{12} + 12\left(\dfrac{1}{6}\right)\left(\dfrac{5}{6}\right)^{11} + 0.29609$

$\qquad\qquad\quad = 0.11216 + 0.26918 + 0.29609 = 0.677 \text{ (3 s.f.)}$

Note:
$p^0 = 1$, $\binom{n}{0} = \binom{n}{n} = 1$ and

$\binom{n}{1} = \binom{n}{n - 1} = n$.

Step 4: Also use the complementary event.

c $P(X \geqslant 1) = 1 - P(X = 0) = 1 - 0.11216 = 0.888 \text{ (3 s.f.)}$

Step 5: Identify the values of Y required.

d $P(5 \leqslant Y \leqslant 7) = P(Y = 5) + P(Y = 6) + P(Y = 7)$

$\qquad\qquad\quad = \binom{12}{5}\left(\dfrac{1}{2}\right)^{12} + \binom{12}{6}\left(\dfrac{1}{2}\right)^{12} + \binom{12}{7}\left(\dfrac{1}{2}\right)^{12}$

$\qquad\qquad\quad = \left(\dfrac{1}{2}\right)^{12} \times (792 + 924 + 792)$

$\qquad\qquad\quad = \left(\dfrac{1}{2}\right)^{12} \times (2508) = 0.612 \text{ (3 s.f.)}$

Note:
When $p = 0.5$, $1 - p = p$.

Step 6: Substitute the identified values into the formula.

Tip:
If your calculator has a binomial distribution function, then get to know how to use it and check these answers.

Example 3.5 At a particular dental practice, the probability that a patient fails to arrive for a dental appointment is 0.16. One morning, a dentist at the practice has 9 appointments. Find the probability that:

    a all patients arrive for their appointments

    b exactly 7 patients arrive for their appointments

    c at least 7 patients arrive for their appointments.

Let X = number of patients who fail to arrive for an appointment. Then $X \sim B(9, 0.16)$.

a $P(X' = 9) = P(X = 0) = (1 - 0.16)^9 = (0.84)^9 = 0.208$ (3 s.f.)

b $P(X' = 7) = P(X = 2) = \binom{9}{2}(0.16)^2(0.84)^7$

$= 36 \times 0.0256 \times 0.29509 = 0.272$ (3 s.f.)

c $P(X' \geqslant 7) = P(X \leqslant 2) = 0.20822 + \binom{9}{1}(0.16)(0.84)^8 + 0.27196$

$= 0.20822 + 0.35694 + 0.27196 = 0.837$ (3 s.f.)

> **Recall:**
> In practice, the outcome with the smaller probability is usually called a 'success' and X' is the complement of X.

SKILLS CHECK **3B: Binomial probability function**

1 A fair six-sided die is rolled 9 times. Calculate the probability of:

    a exactly 1 three;

    b at most 1 three;

    c at least 1 even number;

    d less than 3 numbers divisible by three.

2 A fair coin is tossed 16 times. Calculate the probability of:

    a exactly 10 heads;

    b fewer than 5 heads;

    c more than 6 heads but fewer than 10 heads.

3 Thirteen cards are drawn, with replacement, from a pack of 52 playing cards. Calculate the probability of:

    a at most 1 king;   **b** exactly 4 picture cards;   **c** at least 2 picture cards;

    d between 3 and 5 cards, inclusive, that are either an ace or a picture card.

4 The probability that a particular household receives mail each day is $\frac{5}{6}$ and is independent from one day to the next.

    a Calculate the probability that, during a week (Monday to Saturday), the household receives mail:

        i every day;   **ii** on exactly 5 days;   **iii** on at most 4 days.

    b Calculate the probability that, during a 2-week period, the household receives mail on exactly 10 days.

    c The probability that the household receives junk mail each day is $\frac{1}{9}$ and is independent from one day to the next.

        Calculate the probability that, during a 4-week period, the household receives junk mail on fewer than 3 days.

5 In a particular town, the proportion of households that have satellite TV is 0.36.

    a Calculate the probability that a random sample of 10 households, selected from within the town, contains:

        i exactly 4 households with satellite TV;

        ii at most 2 households with satellite TV;

        iii more than 4 households with satellite TV.

    b **i** Calculate the probability that a sample of 8 households, selected from within the town, contains no households with satellite TV.

        ii What assumption have you made about these 8 households?

        iii Comment, with a reason, on the likely validity of your answer to **i** if the 8 households were all in a small close of expensive bungalows.

6 Within each 1-minute (60 seconds) period, a traffic light shows one of the following four signals for the duration indicated:

Red   20 seconds   Red and Amber   5 seconds   Green   30 seconds   Amber   5 seconds

a A lorry arrives at the traffic light on 6 occasions at random. Calculate the probability that the signal shows Red alone:

  i on exactly 2 occasions;   **ii** on fewer than 2 occasions.

b A van arrives at the traffic light on 12 occasions at random. Calculate the probability that the signal shows Red and Amber on at least 1 occasion.

c A car arrives at the traffic light on 18 occasions at random. Calculate the probability that the signal shows:

  i Red and Amber or Amber on at least 2 occasions;

  ii Green or Amber on exactly 10 occasions.

7 The percentages of black, red, orange, yellow and green wine gums produced by a sweet manufacturer are 16, 19, 23, 28 and 14, respectively.
A tube contains a random sample of 10 wine gums.

a Calculate the probability that the tube contains:

  i exactly 3 that are orange;   **ii** none that are green;

  iii at most 2 that are yellow;   **iv** fewer than 2 that are black;

  v more than 1 but fewer than 4 that are red.

b A tube is found to contain at least one wine gum of each colour. Calculate the probability that this tube contains exactly two that are orange.

 8 A book club circulates an e-mail to all of its subscribers detailing its latest book at a special price. An interested subscriber may purchase the book unseen or send for an inspection copy. A subscriber who sends for an inspection copy may then retain, and so purchase, the book or may choose to return it at no charge. It is known that 13% of subscribers will buy the book unseen and that 25% of subscribers send for an inspection copy, with 80% of these deciding to keep the book.

a A small town has 9 subscribers to the book club. Calculate the probability that, of these 9 subscribers:

  i none buys the book unseen;   **ii** at most 1 buys the book unseen.

b Show that 33% of subscribers actually buy the book. Hence determine the probability that, of the 9 subscribers in the small town:

  i at least 1 buys the book;   **ii** at least 2 but fewer than 5 buy the book.

SKILLS CHECK **3B EXTRA** is on the CD

3.4 Binomial cumulative distribution

Calculation of probabilities using tables.

When a series of binomial probabilities are required, such as $P(5 < X < 15)$, use of the formula is both time consuming and tedious. Fortunately, tables of $P(X \leqslant x)$ have been produced for certain of values of n and p.

> **Note:**
> Lengthy calculations using the formula will **not** be required in the examination.

Example 3.6 In a particular population, 35% of children have black hair, 40% have brown hair, 14% have blond hair and 11% have red hair. A sample of 40 children is selected at random.
Determine the probability that the sample contains:

a at most 15 children with black hair

b more than 15 children with brown hair

c at least 5 children but fewer than 15 children with either blond or red hair

d exactly 16 children with brown hair

e more than 25 children but fewer than 35 children with either black or brown hair

f exactly 5 children with red hair.

Note:
You should refer to Table 1 in the formulae booklet provided in the examination to see the actual values of n and p for which $P(X \leq x)$ is tabulated.

Note:
Any references to Tables 1, 3 and 4 concern the tables in the *AQA Formulae and Statistical Tables* booklet provided in the examination.

Step 1: Identify the values for n and p.

a $n = 40$ and $p = 0.35$ P(black $\leq$ 15) = 0.6946

Note:
3 s.f. answers from the table are acceptable.

Step 2: Identify the value or values of X required.

b $n = 40$ and $p = 0.40$ P(brown > 15) = 1 − P(brown $\leq$ 15)
　　　　　　　　　　　　　　　= 1 − 0.4402 = 0.5598

Tip:
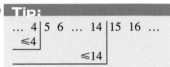

Step 3: Translate into statements about $P(X \leq x)$.

c $n = 40$ and $p = 0.25$ P(5 $\leq$ blond or red < 15)
　　　　　　　　　　　= P(blond or red $\leq$ 14) − P(blond or red $\leq$ 4)
　　　　　　　　　　　= 0.9456 − 0.0160
　　　　　　　　　　　= 0.9296

Tip:
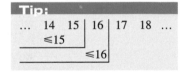

Step 4: Use Table 1 to find the required probability.

d $n = 40$ and $p = 0.40$ P(brown = 16)
　　　　　　　　　　　= P(brown $\leq$ 16) − P(brown $\leq$ 15)
　　　　　　　　　　　= 0.5681 − 0.4402
　　　　　　　　　　　= 0.1279

Tip:
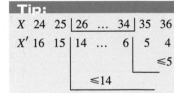

Step 5: As $p > 0.5$, change to the complementary event.

e $n = 40$ and $p = 0.75$

If X denotes 'black or brown' then you require P(25 < X < 35).
As $p = 0.75$ is **not tabulated** but $p = 0.25$ **is tabulated**, change to the complementary event X', then $n = 40$ and $p = 0.25$ and you require

$$P(5 < X' < 15) = P(X' \leq 14) − P(X' \leq 5)$$
$$= 0.9456 − 0.0433 = 0.9023$$

Tip:

X	24	25	26	...	34	35	36
X'	16	15	14	...	6	5	4

≤14
≤5

Step 6: As $p = 0.11$ is not tabulated in Table 1 in the examination booklet, use the formula.

f $n = 40$ and $p = 0.11$

Using the binomial formula

$$P(\text{red} = 5) = \binom{40}{5}(0.11)^5(0.89)^{35}$$

$$= 658008 \times 0.000016105 \times 0.016930 = 0.179 \text{ (3 s.f.)}$$

Note:
Interpolation from the table is **not** accepted in the examination.

Tip:
If your calculator has a binomial cumulative distribution function, then get to know how to use it and check the answers to parts **a** to **e**.

SKILLS CHECK **3C: Binomial cumulative distribution**

1 A fair coin is tossed 25 times. Determine the probability of:

a at most 10 heads;

b more than 15 heads;

c exactly 10 heads;

d more than 10 but fewer than 20 heads;

e more heads than tails.

2 An optician prescribes glasses to 35 per cent of patients following an eye test. Determine the probability that, following an eye test, the number of patients prescribed glasses by the optician, in a random sample of 50 patients:

a is fewer than 20;

b is more than 15;

c is exactly 20;

d is between 15 and 25, inclusive.

3 During 2004 an insurance company received claims from 25 per cent of the motorists that it had insured. For a random sample of 40 motorists, insured with the company in 2004, find the probability that:

a fewer than 15 claimed on their insurance;

b more than 10 claimed on their insurance;

c between 5 and 15, inclusive, claimed on their insurance;

d a majority claimed on their insurance.

4 In a particular professional examination 60 per cent of candidates pass. For a random sample of 30 candidates, find the probability that:

a exactly 15 pass the examination;

b more than 20 pass the examination;

c at least 20 but at most 25 pass the examination;

d fewer than half pass the examination.

5 It is known that 90 per cent of seeds, from a certain variety of petunia, produce plants when sown. A garden nursery grows this variety of petunia plant in plastic strips by sowing exactly 20 seeds in each strip. Determine the probability that a randomly chosen strip produces:

a more than 15 plants;

b fewer than 18 plants;

c exactly 18 plants;

d between 12 and 18, inclusive, plants.

6 Ballpoint pens have black, blue, red or green ink. The proportions of pens with these colours of ink are 0.35, 0.45, 0.15 and 0.05, respectively. Due to a fault in the production process, the four colours of ink were all put into ballpoint pens with identically coloured barrels. Determine the probability that a random sample of 50 of these ballpoint pens contains:

a at most 15 pens with black ink;

b more than 20 pens with blue ink;

c exactly 10 pens with red ink;

d no pens with green ink;

e at least 40 pens with either black or blue ink.

7 Packets of 50 cotton wool balls may be considered to be random samples from a population of balls of which 40% are blue, 40% are pink and 20% are white.

a Determine the probability that a packet contains:

i at most 20 blue balls;

ii at least 25 pink balls;

iii exactly 10 white balls;

iv at least 35 but at most 45 blue or pink balls.

b A packet is known to contain exactly 10 white balls. Determine the probability that the remainder of the packet contains:

i a majority of blue balls;

ii more of one colour than the other;

iii an equal number of blue and pink balls.

8 Each weekday (Monday to Friday) morning at about 10 am, Herbert walks to a nearby garage to buy his morning paper, the Echo. The probability that the garage still has an Echo available when Herbert arrives is 0.70 and is independent from morning to morning.

    a Determine the probability that, over a period of 4 weeks, Herbert obtains an Echo from the garage on:

        i no fewer than 15 mornings;   **ii** at most 10 mornings;

        iii exactly 15 mornings.

    b When the garage has no Echo available, Herbert continues his walk to a shop. The probability that the shop still has an Echo available when Herbert arrives is 0.90 and is independent from morning to morning.

        i Show that the probability that Herbert obtains an Echo each morning is 0.97.

        ii Hence find the probability that, over a period of 8 weeks, Herbert does not obtain an Echo on at least 2 mornings.

SKILLS CHECK **3C EXTRA** is on the CD

3.5 Binomial mean and variance

Mean, variance and standard deviation of a binomial distribution.

If $X \sim B(n, p)$ then:

- mean of X (or μ) $= np$
- variance of X (or σ^2) $= np(1 - p)$
- standard deviation of X (or σ) $= \sqrt{np(1 - p)}$.

When $n > 30$ and $0.1 < p < 0.9$, $P(\mu - 2\sigma < X < \mu + 2\sigma) \approx 95\%$.

Recall:
The mean and variance of a population are denoted by μ and σ^2, respectively.

Note:
Only knowledge and not derivations are required, and the formulae for mean and variance are provided in the examination.

Example 3.7 Paper clips are made in a variety of colours of which 30% are red. Each box of 50 coloured paper clips may be considered to be a random sample.

    a Find values of the mean and standard deviation of the number of red paper clips in a box.

    b Determine the probability that the number of red paper clips in a box is between 9 and 21, inclusive.

Step 1: Substitute the given values of n and p into the formulae for the mean and standard deviation.

    a Here $n = 50$ and $p = 0.3$ so

        mean $= 50 \times 0.3 = 15$

        standard deviation $= \sqrt{50 \times 0.3 \times (1 - 0.3)} = \sqrt{10.5} = 3.24$ (3 s.f.)

Step 2: Translate the required probability into statements about $P(X \le x)$.

Step 3: Use Table 1 in the examination booklet.

    b $P(9 \le \text{red} \le 21) = P(\text{red} \le 21) - P(\text{red} \le 8)$
                     $= 0.9749 - 0.0183 = 0.9566 \ (\approx 95\%)$

Note:
$15 - 2 \times 3.24 = 8.52 \approx 9$.
$15 + 2 \times 3.24 = 21.48 \approx 21$.

Recall:
Use of Table 1.

Example 3.8 Ashley believes that the probability that his mail is delivered before 9 am each day is 0.55 and is independent from day to day.

    a Assuming Ashley's belief is correct, find the mean and standard deviation for the number of times in a week (6 days) that his mail is delivered before 9 am.

b Ashley decides to record each week, for a 13-week period, the number of mornings that his mail is delivered before 9 am. His results are as follows.

$$4 \quad 6 \quad 0 \quad 2 \quad 4 \quad 5 \quad 6 \quad 0 \quad 2 \quad 3 \quad 5 \quad 4 \quad 2$$

i Calculate the mean and standard deviation of these data.

ii Hence comment on the validity of Ashley's belief.

Step 1: Substitute the given values of n and p into the formulae for mean and standard deviation.

Step 2: Calculate the mean and standard deviation of the data.

Step 3: Compare the values obtained.

a Mean $= 6 \times 0.55 = 3.3$
Standard deviation $= \sqrt{6 \times 0.55 \times 0.45} = \sqrt{1.485} = 1.22$ (3 s.f.)

b i Using a calculator, the mean $= 3.31$ (3 s.f.) and the standard deviation $= 2.02$ (3 s.f.).

ii His belief does not appear to be valid since, although the means are almost equal, there is a marked difference in the standard deviations.

> **Note:**
> $\sum x = 43$ and $\sum x^2 = 191$.
> Use of $(\div n)$, giving a s.d. of 1.94, would in similar circumstances gain full marks in the examination.

SKILLS CHECK **3D: Binomial mean and variance**

1 The random variable X has a binomial distribution with $n = 49$ and $p = 0.36$. Find values for the mean, variance and standard deviation of X.

2 A biased die is thrown 3250 times and 650 'sixes' are observed.

a Estimate the probability of a 'six' in a single throw.

b The die is thrown a further 30 times.

i Find the mean and standard deviation for the number of 'sixes' in these 30 throws.

ii Determine the probability of the mean number of 'sixes' in these 30 throws.

 3 An unbiased coin is tossed 40 times and the number of heads, X, is observed.

a Find the mean, μ, and standard deviation, σ, of X.

b i Find $P(X = \mu)$.

ii Find $P(\mu - 2\sigma < X < \mu + 2\sigma)$.

iii Is your answer to **ii** as expected? Justify your answer.

4 A box contains 50 balls of which 10 are white.

a State the mean and variance of the number of white balls observed when 20 balls are randomly selected with replacement.

b Jamil claims to have noted the number of white balls each time he randomly selected 20 balls, with replacement, from the box with the following results.

Number of white balls	0	1	2	3	4	5	6	7	8
Number of occasions	0	4	5	6	10	7	4	3	1

i Calculate the mean and variance of these values.

ii Hence comment on the likely validity of Jamil's claim.

5 Each evening Catherine notes the time that she gets into bed. She believes that the probability that she gets into bed before 11 pm each evening is $\frac{3}{7}$, and is independent from evening to evening.

   a Assuming that Catherine's belief is correct, find the mean and standard deviation of the number of evenings in a week when Catherine gets into bed before 11 pm.

   b During a random sample of 13 weeks, Catherine records, each week, the number of times that she gets into bed before 11 pm with the following results.

      0   4   0   5   3   7   0   6   7   0   5   2   0

      i Calculate the mean and standard deviation of this sample.

      ii Using your results in **a** and **b i**, comment on Catherine's belief.

6 As part of a quality control investigation, Burt took regular samples, of 3 items, from the output of the production line and noted the number of defectives in each sample. His results are given in the table.

Number of defectives	0	1	2	3
Number of samples	75	21	3	1

   a **i** Calculate values for the mean and standard deviation of this sample.

      ii Hence, using your mean value, find an estimate of p, the probability that an item is defective.

   b **i** Assuming that the number of defectives, X, in a sample, can be modelled by B(3, p), find values for the mean and standard deviation of X.

      ii Using your value of p, determine P($X = 0$), P($X = 1$), P($X = 2$) and P($X = 3$).

      iii Hence write down the numbers of samples that you would expect to have 0, 1, 2 and 3 defectives based upon B(3, p).

   c Hence comment on the claim that X can be modelled by B(3, p).

SKILLS CHECK **3D EXTRA** is on the CD

Examination practice Binomial distribution

1 Jeremy sells a magazine which is produced in order to raise money for homeless people. The probability of making a sale is, independently, 0.09 for each person he approaches. Given that he approaches 40 people, find the probability that he will make:

   a 2 or fewer sales;   **b** exactly 4 sales;   **c** more than 5 sales.   [AQA(B) Jan 2002]

2 A gardener plants beetroot seeds. The probability of a seed not germinating is 0.35, independently for each seed.

Find the probability that, in a row of 40 seeds, the number not germinating is:

   a 9 or fewer;   **b** 7 or more;   **c** equal to the number germinating.

                                                    [AQA(B) May 2003]

3 A specialist bicycle shop builds made-to-measure bicycle frames. The shop has 40 orders for frames. Past experience shows that the probability of a customer, who has ordered a frame, failing to complete the purchase is 0.04 and is independent for each customer.

   a Find the probability that, of the 40 orders, the number of purchases not completed is:

      i 4 or fewer;   **ii** exactly 2.

   b Find the probability that the purchase is completed for all 40 orders.   [AQA(B) Nov 2002]

4 The percentages of black, red, orange, green and yellow jelly babies produced by a sweet manufacturer are 20, 21, 30, 15 and 14, respectively.

Determine the probability that a random sample of 15 jelly babies, from those produced by the sweet manufacturer, contains

a at least 5 that are black,

b more than 2 but fewer than 7 that are orange,

c at most 2 that are yellow. [AQA(NEAB) Feb 2001]

Please note that this question is NOT from the live examinations for the current specification.

5 A manufacturer of balloons produces 40 per cent that are oval and 60 per cent that are round.

Packets of 20 balloons may be assumed to contain random samples of balloons. Determine the probability that such a packet contains:

a an equal number of oval balloons and round balloons;

b fewer oval balloons than round balloons.

A customer selects packets of 20 balloons at random from a large consignment until she finds a packet with exactly 12 round balloons.

c Give a reason why a binomial distribution is **not** an appropriate model for the number of packets selected. [AQA(A) June 2002]

6 Twenty per cent of coloured beads used in costume jewellery are blue.

a Determine the probability that
i in a string of 20 beads, more than 3 beads are blue,
ii in a string of 28 beads, exactly 4 beads are blue.

b State **one** assumption that you have made about the beads in answering part **a**.

c A shop has a display of 10 strings each consisting of a random selection of 28 beads. A bead is selected at random from each string.

Explain why a binomial distribution is unlikely to be a suitable model for the number of blue beads in the sample of 10 beads. [AQA(A) June 2001]

7 A bin contains 400 coloured erasers that fit on the ends of pencils. The number of erasers of each colour is as follows.

Colour	Green	Blue	Red	Yellow
Number	88	60	160	92

a A random sample of 25 erasers is selected, **with replacement**, from the bin. Find the probability that:
i exactly 2 erasers are green;
ii at most 3 erasers are blue;
iii between 8 erasers and 12 erasers, inclusive, are red.

b Erasers are selected at random, **without replacement**, from the bin until 5 yellow erasers are obtained.

Give **two** reasons why a binomial distribution does **not** model the number of erasers selected.
[AQA(A) June 2004]

8 Eight friends take a picnic to a cricket match. As her contribution to the picnic, Hilda buys eight sandwiches at a supermarket. She selects the sandwiches at random from those on display. The probability that a sandwich is suitable for vegetarians is independently 0.3 for each sandwich.

a Find the probability that, of the eight sandwiches, the number suitable for vegetarians is:

i 2 or fewer;

ii exactly 2;

iii more than 3.

b Two of the eight friends are vegetarians. Hilda decides to ensure that the eight sandwiches she takes to the match will include at least two suitable for vegetarians. If, having selected eight sandwiches at random, she finds they include fewer than two suitable for vegetarians she will replace one, or if necessary two, of the sandwiches unsuitable for vegetarians with the appropriate number of sandwiches suitable for vegetarians.

State whether or not the binomial distribution provides an appropriate model for the number of sandwiches suitable for vegetarians which Hilda takes to the match. Explain your answer.

c In fact the eight sandwiches which Hilda took to the match contained four suitable for vegetarians. The first four friends to eat a sandwich were not vegetarians. Each selected one of the available sandwiches at random and ate it.

State whether or not the binomial distribution provides an appropriate model for the number of sandwiches suitable for vegetarians eaten by these four friends. Explain your answer.

[AQA(B) Jan 2004]

9 Gerhard walks to school each morning. The probability that he arrives late is 0.15 and is independent of whether he arrives late on any other morning.

a Find, for a week when he walks to school on five mornings:

i the probability he arrives late on two or fewer mornings;

ii the probability he arrives late on more than three mornings;

iii the mean and the standard deviation of the number of mornings on which he arrives late.

b The following table summarises the number of late arrivals of all pupils who attended Gerhard's school on five mornings of a particular week.

Number of late arrivals	Number of pupils
0	275
1	111
2	33
3	12
4	13
5	16

i Calculate the mean and the standard deviation of the data in the table.

ii Show that, for the pupils represented in the table, an estimate of the probability of a pupil arriving late on a particular morning is 0.15.

c It is suggested that the data in the table could be modelled by a binomial distribution.

i Comment on this suggestion in the light of your calculations in parts **a** and **b**.

ii In the context of this question, give **two** possible reasons why the binomial distribution may **not** be a suitable model for the data in part **b**.

[AQA(B) May 2004]

4 Normal distribution

4.1 Continuous random variables

Continuous random variables.

A random variable is said to be **continuous** when a list of its possible values cannot be constructed, although a range for its possible values may well be available. Continuous random variables arise from **measuring** a characteristic such as time, height, length, area, volume and weight.

Recall:
Discrete random variables.

As there is no finite list of possible values for a continuous random variable, the probability of any exact value is zero, i.e.

$$P(X = x) = 0 \text{ for all } x$$

Note:
In practice, the number of possible values may be limited simply by the precision of the measuring device.

However, non-zero probabilities can be found for ranges of values of a continuous random variable,

$$P(a < X < b) = P(a \leqslant X \leqslant b) > 0 \text{ for } a < b$$

Note:
For continuous random variables, $< \equiv \leqslant$ and $> \equiv \geqslant$ because $P(X = x) = 0$.

The (probability) **distribution** for a continuous random variable, X, is defined by the range of possible values of X, together with a function, $f(x)$, called a **probability density function**, that describes a curve such that areas under it represent probabilities. Clearly, for the range of possible values, the total area under the curve must be 1 and the curve cannot fall below the horizontal axis.

Note:
Often shortened to PDF or pdf.

For most continuous distributions it is possible to find, from f(x), the **cumulative distribution function**, $F(x)$, such at $F(x) = P(X \leqslant x)$.

Note:
Often shortened to CDF or cdf.

Probabilities for a continuous random variable may be found from its probability density function or from its cumulative distribution function.

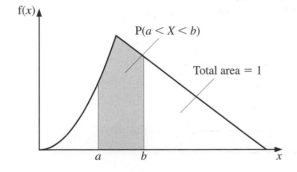
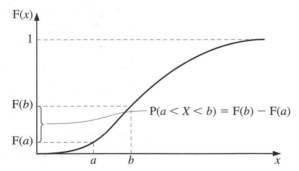

4.2 Properties of normal distributions

Properties of normal distributions.

The normal distribution is by far the most useful continuous distribution. It can be used to describe many situations where repeated measurements are made of a characteristic that tends to have relatively few very small or very large values, but a relatively large number of values about a middle value.

Typical characteristics are:
- the time taken to complete a puzzle
- the length of string on a reel
- the volume of paint in a tin
- the weight of potatoes in a bag.

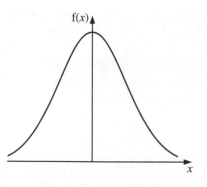

If the random variable X has a normal distribution, then the graph of $f(x)$ is a curve that is **bell-shaped and symmetrical**, and such that the **area under it from $-\infty$ to $+\infty$ is 1**.

A normal distribution has **two parameters**: μ, the **mean**; and σ^2, the **variance**. The expression

$$X \sim N(\mu, \sigma^2)$$

is read as

> 'The (continuous) random variable, X, has a normal distribution with mean μ and variance σ^2.'

Recall:
The parameters of a binomial distribution.

Note:
The equivalent of this statement is provided in the examination.

Changes in the value of μ change the location of the curve on the x-axis, whereas changes in the value of σ change the shape of the curve, although it always remains bell-shaped and symmetrical.

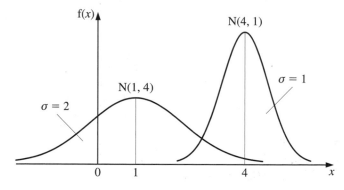

If $X \sim N(\mu, \sigma^2)$, then:
- the curve is symmetrical about μ
- μ is also the mode and the median
- approximately two-thirds of the values lie within $\mu \pm \sigma$
- approximately 95% of the values lie within $\mu \pm 2\sigma$
- more than 99% of the values lie within $\mu \pm 3\sigma$.

4.3 Calculation of probabilities

Calculation of probabilities.

If $X \sim N(\mu, \sigma^2)$, then the actual form of its probability density function is given by

$$f(x) = \frac{1}{\sigma\sqrt{2\pi}}\, e^{-\frac{1}{2}\left(\frac{x-\mu}{\sigma}\right)^2}$$

Note:
Although this expression is provided in the examination, it will **not** be required.

Finding areas under this curve requires numerical techniques. However, this difficulty has been overcome by the provision of tables of the cumulative distribution function for the **standardised normal random variable Z**, which has $\mu = 0$ and $\sigma = 1$, i.e. $Z \sim N(0, 1)$.

In other words, such tables provide values of $P(Z \leq z)$ for values of z, where $Z \sim N(0, 1)$.

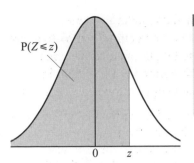

$P(Z \leq z)$

Note:

You should refer to Table 3 in the formulae booklet provided in the examination to see the actual values of z for which $P(Z \leq z)$ is tabulated.

Example 4.1 The random variable Z has a standardised normal distribution. Determine the probability that a value of Z:

a is less than 1.25

b is greater than 0.82

c is at least -1.47

d is at most -0.43

e is between 0.32 and 1.95

f is between -2.16 and 0.87

g is at least -2.44 but at most -1.90.

Recall:

Since Z is a continuous random variable,
$P(Z = z) = 0$ so
$P(Z \leq z) = P(Z < z)$ and
$P(Z \geq z) = P(Z > z)$ etc.

Note:

Sketches of the PDF for Z often help in the use of Table 3. In particular, as to whether the required area is 'small' or 'large'.

Step 1: Sketch a bell-shaped curve, centred on 0.

Step 2: Mark on the z-axis the required value(s) of Z.

Step 3: Shade the area required and note whether it is 'small' or 'large'.

Step 4: Use Table 3 in the examination booklet to find the required probability, checking as to whether it agrees with your note of its relative size,

a $P(Z < 1.25) = 0.89435$ or 0.894 (3 s.f.)

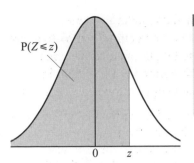

Recall:

The curve for the PDF of Z is symmetrical about zero.

b $P(Z > 0.82) = 1 - P(Z < 0.82)$
$\qquad = 1 - 0.79389$
$\qquad = 0.20611$ or 0.206 (3 s.f.)

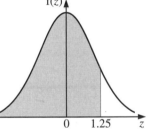

Note:

$P(Z > z) = 1 - P(Z < z)$.

c $P(Z > -1.47) = P(Z < 1.47)$
$\qquad = 0.92922$ or 0.922 (3 s.f.)

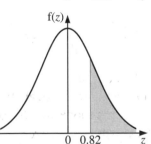

Note:

$P(Z > -z) = P(Z < z)$.

d $P(Z < -0.43) = P(Z > 0.43)$
$\qquad = 1 - P(Z < 0.43)$
$\qquad = 1 - 0.66640$
$\qquad = 0.3336$ or 0.334 (3 s.f.)

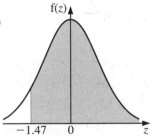

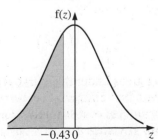

Note:

$P(Z < -z) = 1 - P(Z < z)$.

e $P(0.32 < Z < 1.95)$
$= P(Z < 1.95) - P(Z < 0.32)$
$= 0.97441 - 0.62552$
$= 0.34889$ or 0.349 (3 s.f.)

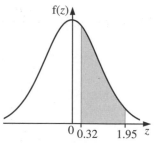

Note:

$P(z_1 < Z < z_2) =$
$P(Z < z_2) - P(Z < z_1)$.

f $P(-2.16 < Z < 0.87)$
$= P(Z < 0.87) - P(Z < -2.16)$
$= P(Z < 0.87) - (1 - P(Z < 2.16))$
$= 0.80785 - (1 - 0.98461)$
$= 0.80785 - 1 + 0.98461$
$= 0.79246$ or 0.792 (3 s.f.)

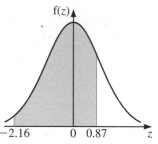

Recall:

Notes from **e** and **d**.

g $P(-2.44 < Z < -0.90)$
$= P(0.90 < Z < 2.44)$
$= P(Z < 2.44) - P(Z < 0.90)$
$= 0.99266 - 0.81594$
$= 0.17672$ or 0.177 (3 s.f.)

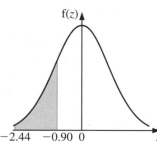

Recall:

Symmetry about 0 and note from **d**.

Tip:

If your calculator has a standardised normal distribution function, then get to know how to use it and check these answers.

If $X \sim N(\mu, \sigma^2)$, then values of X can be **standardised** (linearly scaled) to values of $Z \sim N(0, 1)$ by the transformation

$$Z = \frac{X - \mu}{\sigma}$$

Note:

The divisor is σ, not σ^2.

Example 4.2 The weight of salt, delivered by a machine into cardboard tubs, may be assumed to be a normal random variable with mean 355 grams and standard deviation 2 grams. Determine the probability that the weight of salt in a randomly selected tub is:

a less than 350 grams

b more than 355 grams

c between 354 grams and 358 grams

d within one standard deviation of the mean

e in excess of 0.5% of the mean weight above the mean weight.

Note:

Your calculator may have a normal distribution function that removes the need for standardising. However, no calculator that is permitted in the examination can solve all of the problems in Section 4.4. You must therefore know how to standardise variables and how to use the supplied tables.

Step 1: Write down the actual form for standardising.

Let X be the weight, in grams, of salt delivered by the machine.
Then $X \sim N(355, 2^2)$ and so $Z = \dfrac{X - 355}{2}$

Step 2: Standardise the required value(s) of X to values of Z.

Step 3: Use Table 3 in the examination booklet to find the required probability (checking its 'size').

a $P(X < 350) = P\left(Z < \dfrac{350 - 355}{2}\right) = P(Z < -2.5)$

$= 1 - P(Z < 2.5) = 1 - 0.97725$
$= 0.02275$ or 0.0228 (3 s.f.)

b $P(X > 355) = 0.5$ since X is symmetrical about 355

Note:

Do **not** make the mistake, here and elsewhere, of replacing, for example, (<350) with ($\leqslant 349.9$) or ($\leqslant 349.5$) or ($\leqslant 349$).

c
$$P(354 < X < 358) = P\left(\frac{354 - 355}{2} < Z < \frac{358 - 355}{2}\right)$$

$$= P(-0.5 < Z < 1.5)$$
$$= P(Z < 1.5) - P(Z < -0.5)$$
$$= P(Z < 1.5) - (1 - P(Z < 0.5))$$
$$= 0.93319 - (1 - 0.69146)$$
$$= 0.93319 - 1 + 0.69146$$
$$= 0.62465 \text{ or } 0.625 \text{ (3 s.f.)}$$

Tip:

If in doubt, sketch a bell-shaped curve, centred on μ. Mark on the x-axis the required value(s) of X. Shade the area required and note whether it is 'small' or 'large'.

d The probability that X is within one standard deviation (2) of the mean (355) is given by

$$P(353 < X < 357) = P\left(\frac{353 - 355}{2} < Z < \frac{357 - 355}{2}\right)$$
$$= P(-1 < Z < 1)$$

Recall:

Approximately two-thirds of the values lie within $\mu \pm \sigma$.

which is the probability that Z is within one standard deviation (1) of the mean (0).
$$P(-1 < Z < 1) = P(Z < 1) - P(Z < -1)$$
$$= P(Z < 1) - (1 - P(Z < 1))$$
$$= 0.84134 - (1 - 0.84134)$$
$$= 0.84134 - 1 + 0.84134$$
$$= 0.68268 \text{ or } 0.683 \text{ (3 s.f.)}$$

Tip:

Find probabilities for $\mu \pm 2\sigma$ and for $\mu \pm 3\sigma$ and compare with 4th and 5th listed properties on page 45.

e The mean weight is 355 grams so 0.5% is given by

$$\frac{0.5}{100} \times 355 = 1.775$$

$$P(X > (355 + 1.775)) = P(X > 356.775) = P\left(Z > \frac{356.775 - 355}{2}\right)$$
$$= P(Z > 0.8875) \approx P(Z > 0.89)$$
$$= 0.81327 \text{ or } 0.813 \text{ (3 s.f.)}$$

Note:

Although the use of linear interpolation in Table 3 is valid, it is **not** expected in the examination. Values of z may thus be rounded to 2 d.p. when using Table 3.

SKILLS CHECK **4A: Calculation of probabilities**

1 Given that $Z \sim N(0, 1)$, determine:

a $P(Z = 0.5)$ **b** $P(Z < 0.86)$ **c** $P(Z > 1.35)$ **d** $P(Z \geqslant -0.32)$ **e** $P(Z \leqslant -1.27)$

2 Given that $Z \sim N(0, 1)$, determine:

a $P(0.66 < Z < 1.33)$ **b** $P(0 < Z < 2.06)$

c $P(-1.44 < Z < 0.92)$ **d** $P(-1.73 \leqslant Z \leqslant -0.17)$

3 Given that $X \sim N(20, 6.25)$, determine:

a $P(X < 22)$ **b** $P(X > 24)$ **c** $P(X = 22)$ **d** $P(X > 19)$ **e** $P(X < 17)$

4 Given that $Y \sim N(153, 100)$, determine:

a $P(155 < Y < 170)$ **b** $P(140 < Y < 160)$ **c** $P(143 < Y < 153)$ **d** $P(130 \leqslant Y \leqslant 145)$

5 The volume of path cleaning solution in 20-litre containers may be assumed to be normally distributed with mean 20.12 litres and standard deviation 0.06 litres. Find the probability that the volume of the solution in a randomly chosen container is:

a less than 20.2 litres, **b** more than 20 litres,

c between 20.1 litres and 20.2 litres, **d** within $1\frac{1}{2}$ standard deviations of the mean volume.

6 The weight of 25-kg bags of golden gravel may be assumed to be normally distributed with a mean of 26.25 kg and a standard deviation of 0.9 kg. Determine the probability that the weight of a randomly selected bag of golden gravel is:

    a less than 26.5 kg,   **b** more than 25 kg,

    c between 26 kg and 27 kg,   **d** more than 25 kg but less than 26 kg,

    e exactly 1 kg overweight,

    f overweight by more than 5% of the advertised weight.

7 The time taken by Carlos to walk from home into town may be modelled by a normal random variable with a mean of 21 minutes and a standard deviation of 4 minutes.

    a Determine the probability that the time taken by Carlos to walk from home into town on a randomly chosen day is:

        i less than 25 minutes,   **ii** between 15 minutes and 20 minutes.

    b Carlos arranges to meet his girlfriend, Bianca, in town at 7.15 pm. He leaves home at 6.50 pm.

        i Find the probability that he is late for his meeting with Bianca.

        ii Bianca leaves their meeting place 5 minutes after the arranged time. Find the probability that he is late for the meeting but Bianca has not left.

8 A machine cuts plastic strips for 30 cm rulers. The lengths of cut strips may be modelled by a normal random variable with mean 31.5 cm and standard deviation 0.15 cm.

    a Determine the probability that a randomly chosen strip is:

        i less than 31.75 cm,   **ii** more than 31.25 cm,   **iii** between 31.2 cm and 31.7 cm.

    b Given that there must be at least 0.6 cm unused at each end of each ruler, determine the probability that a randomly chosen strip is of insufficient length.

9 Liam is a member of an athletics club. In long jump competitions, the lengths of his jumps are normally distributed with a mean of 7.65 m and a standard deviation of 0.18 m.

    a Calculate the probability of him jumping:

        i more than 8 m,   **ii** less than 7.50 m,   **iii** between 7.50 m and 7.75 m.

    b Mark also belongs to the same athletics club. In long jump competitions, the lengths of his jumps are normally distributed with a mean of 7.45 m and a standard deviation of 0.32 m. The athletics club has to select either Liam or Mark to be its long jump competitor at a major athletics meeting. In order to qualify for the final rounds of long jumps at the meeting, it is necessary to achieve a jump of at least 8 m in the qualifying rounds. State, with justification, which of the two athletes should be selected.

10 A house builder orders double-glazed units for some of the windows of houses on a large new estate. Each unit should nominally be 60 cm long and 50 cm wide. However, the actual length, L cm, may be assumed to be a normal random variable with mean 59.8 and standard deviation 0.3, and the actual width, W cm, may also be assumed to be a normal random variable but with mean 50.1 and standard deviation 0.3.

    a Determine:

        i $P(L > 60)$ **ii** $P(W < 49.5)$ **iii** $P(59.0 < L < 60.5)$ **iv** $P(49.0 < W < 50.5)$

    b Units are satisfactory when both $59.0 < L < 60.5$ and $49.0 < W < 50.5$. Given that L and W are independent and that 2000 units are delivered onto the estate, how many units would you expect to be unsatisfactory? Give your answer to the nearest 10 units.

SKILLS CHECK **4A EXTRA** is on the CD

Sometimes, given a probability p, it is necessary to find the corresponding value z such that $P(Z \leq z) = p$ where $Z \sim N(0, 1)$.

Tables providing values of z corresponding to given values of p are available. In effect, such tables are the 'reverse' of those used in Section 4.3, which provide values of p for given values of z.

Note:

You should refer to Table 4 in the formulae booklet provided in the examination to see the actual values of p for which values of z, satisfying $P(Z \leq z) = p$, are tabulated.

Example 4.3 The random variable Z has a standardised normal distribution. Find the value of z such that:

a $P(Z < z) = 0.88$

b $P(Z > z) = 0.44$

c $P(Z > z) = 0.956$

d $P(Z < z) = 0.025$

e $P(-z < Z < z) = 0.98$.

Step 1: Sketch a bell-shaped curve, centred on 0.

Step 2: Shade the given area and mark on the z-axis the corresponding value of z, noting whether it is positive or negative.

Step 3: If $p < 0.5$, work with $1 - p$.

Step 4: Use Table 4 in the examination booklet to find the required value of z, checking as to whether it agrees with your note of its sign.

a $P(Z < z) = 0.88$ implies, directly from Table 4 in the examination booklet, that $z = 1.1750$.

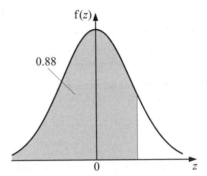

b $P(Z > z) = 0.44$ implies that

$P(Z < z) = 1 - 0.44 = 0.56$

so $z = 0.1510$.

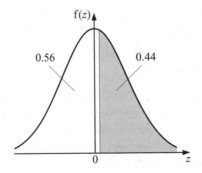

c $P(Z > z) = 0.956$ implies that z must be negative since $P(Z > 0) = 0.5$.
Using Table 4 gives

$P(Z < 1.7060) = 0.956$

Thus, by symmetry, $z = -1.7060$.

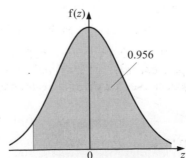

Recall:

The curve for the PDF of Z is symmetrical about zero.

d $P(Z < z) = 0.025$ implies that z must be negative since $P(Z < 0) = 0.5$.
As $0.025 < 0.5$, then working with $1 - 0.025 = 0.975$ and using Table 4 gives

$P(Z < 1.9600) = 0.975$

Thus, by symmetry, $z = -1.9600$.

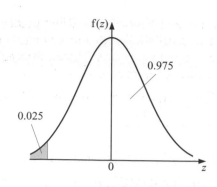

e $P(-z < Z < z) = 0.98$ implies, because of symmetry, that

$$P(Z > z) = \frac{1 - 0.98}{2} = 0.01$$

so $P(Z < z) = 0.98 + 0.01 = 0.99$

giving $z = 2.3263$.

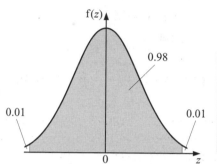

> **Tip:**
> If your calculator has a facility that provides values of z corresponding to given values of p, then get to know how to use it and check these answers.

Example 4.4 The times taken by nine-year-old children to complete a particular mathematical exercise may be assumed to be normally distributed with a mean of 12.6 minutes and a standard deviation of 1.5 minutes. Find the time exceeded by 90% of children.

Step 1: Write down the actual form for standardising.

Let X denote the time, in minutes, to complete the exercise, then $X \sim N(12.8, 1.5^2)$ and so $Z = \dfrac{X - 12.8}{1.5}$

Step 2: Use Table 4 to find the required z-value (checking its 'sign').

A value of x is required such that $P(X > x) = 0.90$ (90%).

Step 3: Equate the standardised x-value to the z-value and solve for x.

However, $P(Z < 1.2816) = 0.90$ so, by symmetry,

$P(Z > -1.2816) = 0.90$

Thus $\dfrac{x - 12.8}{1.5} = -1.2816$ so $x = 10.9$ (3 s.f.).

The time exceeded by 90% of children is 10.9 minutes.

> **Tip:**
> If in doubt, sketch a N(0, 1) curve and shade the area required. Note whether the corresponding z-value is positive or negative.

Example 4.5 Lengths of 2-metre wooden boards cut by a machine are known to be normally distributed with a standard deviation of 3.4 centimetres. Given that 5% of the boards have lengths in excess of 2.05 metres, find, to the nearest centimetre, the mean length of the boards.

Step 1: Write down the actual form for standardising.

Let X denote the length, **in centimetres**, of a board, then $X \sim N(\mu, 3.4^2)$ and so $Z = \dfrac{X - \mu}{3.4}$

A value of μ is required such that $P(X > 205) = 0.05$ (5%).

Step 2: Use Table 4 to find the required z-value (checking its 'sign').

However, $P(Z < 1.6449) = 0.95$, so $P(Z > 1.6449) = 1 - 0.95 = 0.05$.

Step 3: Equate the standardised x-value to the z-value and solve for μ.

Thus $\dfrac{205 - \mu}{3.4} = 1.6449$, so $\mu = 199$.

The mean length of the boards is 199 centimetres or 1.99 metres.

> **Tip:**
> Always use a common unit of measurement.

> **Tip:**
> If the probability given (p) is less than 0.5, then first consider $1 - p$.

Example 4.6 The volume of lemonade in 2-litre bottles may be assumed to be normally distributed with a mean of 2.04 litres. Given that 98% of bottles contain more than 2 litres, determine the standard deviation of the volume of lemonade.

Step 1: Write down the actual form for standardising.

Let X denote the volume, in litres, of lemonade, then $X \sim N(2.04, \sigma^2)$ and so $Z = \dfrac{X - 2.04}{\sigma}$.

A value of σ is required such that $P(X > 2) = 0.98$ (98%).

Step 2: Use Table 4 to find the required z-value (checking its 'sign').

However, $P(Z < 2.0537) = 0.98$, so, by symmetry,

$P(Z > -2.0537) = 0.98$

Step 3: Equate the standardised x-value to the z-value and solve for σ.

Thus $\dfrac{2 - 2.04}{\sigma} = -2.0537$, so

$\sigma = 0.0195$ litres (3 s.f.) or 19.5 millilitres.

> **Recall:**
> σ cannot be negative!

Example 4.7 A utility company finds that 8% of its domestic customers with water meters use more than 250 cubic metres of water in a year, and that 15% of such customers use less than 150 cubic metres of water in a year. Determine the mean and standard deviation for the volume of water used by the company's domestic customers in a year.

Step 1: Write down the general form for standardising.

Let X denote the volume, in cubic metres, of water used in a year, then $X \sim N(\mu, \sigma^2)$ and so $Z = \dfrac{X - \mu}{\sigma}$

Values of μ and σ are required such that $P(X > 250) = 0.08$ (8%) and $P(X < 150) = 0.15$ (15%).

Step 2: Use Table 4 to find the required z-values (checking their 'signs').

However, $P(Z < 1.4051) = 0.92$, so $P(Z > 1.4051) = 1 - 0.92 = 0.08$.

Also, $P(Z < 1.0364) = 0.85 = 1 - 0.15$, so $P(Z > 1.0364) = 0.15$.

Thus, by symmetry, $P(Z < -1.0364) = 0.15$.

Step 3: Equate the standardised x-values to the z-values and solve the resultant simultaneous equations for μ and σ.

Hence $\dfrac{250 - \mu}{\sigma} = 1.4051$ or $250 - \mu = 1.4051\sigma$ ①

and $\dfrac{150 - \mu}{\sigma} = -1.0364$ or $150 - \mu = -1.0364\sigma$ ②

Subtracting equation ② from equation ① gives

$100 = 2.4415\sigma$,   so $\sigma = 40.96$

Substituting for σ in equation ① gives

$250 - \mu = 1.4051 \times 40.96 = 57.55$,   so $\mu = 192.4$

Thus, to 3 s.f., the mean is 192 cubic metres and the standard deviation is 41.0 cubic metres.

> **Recall:**
> Solving a pair of linear simultaneous equations from Core 1.

SKILLS CHECK **4B: Calculation of values. means and standard deviations**

1 Given that $Z \sim N(0, 1)$, find the value z such that:

a $P(Z < z) = 0.75$ **b** $P(Z < z) = 0.975$ **c** $P(Z > z) = 0.33$

d $P(Z > z) = 0.98$ **e** $P(Z < z) = 0.15$

2 Given that $Z \sim N(0, 1)$, find the value z such that:

a $P(-z < Z < z) = 0.66$ **b** $P(-z < Z < z) = 0.97$

c $P(-z < Z < z) = 0.50$

3 Given that $X \sim N(20, 6.25)$, find the value x such that:

a $P(X < x) = 0.80$ **b** $P(X > x) = 0.25$

c $P(X > x) = 0.70$ **d** $P(X < x) = 0.15$

4 Given that $Y \sim N(153, 100)$, find values of y_1 and y_2, symmetrical about the mean, such that:

a $P(y_1 < Y < y_2) = 0.80$ **b** $P(y_1 < Y < y_2) = 0.40$

 5 The diameters of a particular variety of mass-produced pizza are normally distributed with a mean of 30 cm and a standard deviation of 2 cm.

a Find the diameter exceeded by:

  i 10 per cent of pizzas,   **ii** 95 per cent of pizzas.

b Find two values, symmetrical about the mean, within which lie the diameters of 95 per cent of pizzas.

6 The weights of fruit allsorts in a packet may be modelled by a normal random variable with standard deviation 4.5 grams. Given that 98% of such packets contain at least 215 grams, calculate the mean weight of fruit allsorts in a packet.

7 The volume of car shampoo in 5-litre plastic containers is normally distributed with a mean of 5.2 litres.

a Given that at most 1% of containers should contain less than 5 litres, find the maximum value for the standard deviation of the volume of car shampoo in a container.

b The minimum value possible for this standard deviation is 0.10 litres. Given that the value of the standard deviation is reduced to this minimum value, by how much should the mean volume be increased so that at most 1% of containers should contain less than 5 litres? Give your answer to the nearest 0.01 litres.

8 A builders' merchant supplies soft sand in 1000 kg sacks. A mechanical shovel, which incorporates an automatic weighing device, fills these sacks. An analysis of the weights of sand in a sample of sacks shows that 98% contain less than 1075 kg, and that 60% contain less than 1025 kg. Assuming that the weight of sand in a sack is normally distributed, calculate estimates for the mean and standard deviation of the weight of sand in a sack.

 9 A wholesaler buys cauliflowers from a farmer for distribution to retail greengrocers. The weights of the cauliflowers can be modelled by a normal distribution with a mean of 640 grams and a standard deviation of 160 grams. The wholesaler classifies the lightest 15% of cauliflowers as small, the heaviest 25% as large, and the remainder as medium.

a Calculate the weight that a cauliflower

  i must exceed to be classified as large,   **ii** must fall below to be classified as small.

b The wholesaler decides to classify cauliflowers with weights above 850 grams as extra large.

  i Find the percentage of extra large cauliflowers.

  ii Find, to the nearest 10 grams, the median weight of extra large cauliflowers.

10 A small manufacturing firm has 3 machines, *U*, *V* and *W*, that produce ball bearings. The diameters of the ball bearings, produced by each of these three machines, may be assumed to be normally distributed. The firm rejects, as undersize, all ball bearings with diameters less than 9.490 mm, and rejects, as oversize, all ball bearings with diameters greater than 9.520 mm.

   a Bearings produced on Machine *U* have diameters with mean 9.505 mm and standard deviation 0.009 mm. Determine

       i the percentage of ball bearings produced on this machine that are rejected as oversize,

       ii the percentage of ball bearings produced on this machine that are acceptable.

   b Machine *V* produces ball bearings that have a mean diameter of 9.504 mm, of which 2.5% are rejected as undersize. Calculate the standard deviation of the diameters of ball bearings produced on Machine *V*.

   c Of the ball bearings produced on Machine *W*, 0.5% are rejected as undersize, and 3.5% are rejected as oversize. Calculate the mean and standard deviation of the diameters of ball bearings produced on Machine *W*.

SKILLS CHECK **4B EXTRA** is on the CD

Examination practice   Normal distribution

1 The content, in milligrams, of vitamin C in a litre carton of cranberry juice can be modelled by a normal distribution with a mean of 32 and a standard deviation of 2.

   a Determine the probability that, for a carton chosen at random, the vitamin C content is less than 30 mg.

   b Find, to the nearest milligram, the value of the mean required to ensure that the percentage of cartons with a vitamin C content of less than 30 mg is 2.5.   [AQA(A) Nov 2002]

2 The weight of a particular variety of orange is normally distributed with a mean of 155 grams and a standard deviation of 10 grams.

   a Determine the proportion of oranges with weights between 145 grams and 165 grams.

   b Determine the weight exceeded by 67 per cent of the oranges.   [AQA(NEAB) March 1999]

   Please note that this question is NOT from the live examinations for the current specification.

3 The weights of bags of red gravel may be modelled by a normal distribution with mean 25.8 kg and standard deviation 0.5 kg.

   a Determine the probability that a randomly selected bag of red gravel will weigh:

       i less than 25 kg;

       ii between 25.5 kg and 26.5 kg.

   b Determine, to two decimal places, the weight exceeded by 75% of bags.   [AQA(A) Jan 2003]

4 The weights of hens' eggs are normally distributed with a mean of 65 grams and a standard deviation of 10 grams.

   Eggs whose weights are less than 56.5 grams are classified as small.

   a Calculate the proportion of eggs classified as small.

   Those eggs which are not classified as small are classified as medium or large in equal proportion.

   b Calculate, to one decimal place, the weight above which eggs are classified as large.

   [AQA(NEAB) June 1999]

   Please note that this question is NOT from the live examinations for the current specification.

5 The volume, L litres, of emulsion paint in a plastic tub may be assumed to be normally distributed with mean 10.25 and variance σ^2.

   a Assuming that $\sigma^2 = 0.04$, determine $P(L < 10)$.

   b Find the value of σ so that 98% of tubs contain more than 10 litres of emulsion paint.

[AQA(A) June 2004]

 6 A machine dispenses peanuts into bags so that the weight of peanuts in a bag is normally distributed.

   a Initially the mean weight of peanuts in a bag is 128.5 g and the standard deviation is 1.5 g. Find the probability that the weight of peanuts in a randomly chosen bag exceeds 130 g.

   b The machine is given a minor overhaul that changes the mean weight, μ, of peanuts in a bag without affecting the standard deviation. Following the overhaul, 14% of bags contain more than 130 g of peanuts. Find, to four significant figures, the new value for μ.

   c Later the machine requires a major repair, following which the mean weight of peanuts in a bag is 128.3 g, and 4% of bags contain less than 126 g. Find, to three significant figures, the standard deviation of the weight of peanuts in a bag after this major repair.   [AQA(NEAB) June 1997]

   Please note that this question is NOT from the live examinations for the current specification.

7 A teacher travels from home to work by car each weekday by one of two routes, X or Y.

   a For route X, her journey times are normally distributed with a mean of 30.4 minutes and a standard deviation of 3.6 minutes.

     i Calculate the probability that her journey time on a particular day takes between 25 minutes and 35 minutes.

     ii Determine the 80th percentile of her journey times.

   b For route Y, her journey times are normally distributed with a mean of 33.0 minutes, and 30 per cent of journeys take between 33 minutes and 35 minutes.

   Determine, to one decimal place, the standard deviation of these journey times.

   c The teacher is required to arrive at work no later than 8.45 am.

   On a day when she leaves home at 8.10 am, determine which of routes X or Y she should use.

[AQA(NEAB) March 2000]

   Please note that this question is NOT from the live examinations for the current specification.

8 a The time, X minutes, taken by Fred Fast to install a satellite dish may be assumed to be a normal random variable with mean 134 and standard deviation 16.

     i Determine $P(X < 150)$.

     ii Determine, to one decimal place, the time exceeded by 10 per cent of installations.

   b The time, Y minutes, taken by Sid Slow to install a satellite dish may also be assumed to be a normal random variable, but with

     $P(Y < 170) = 0.14$ and $P(Y > 200) = 0.03$.

   Determine, to the nearest minute, values for the mean and standard deviation of Y.

[AQA(A) June 2003]

9 In order not to be late for a job interview, Anita needs to leave her house in a taxi no later than 3.00 pm. Past experience has shown that, when she telephones for a taxi from company A, the time it takes to arrive at her house may be modelled by a normal distribution with a mean of 12 minutes and a standard deviation of 3 minutes.

   a Give that she telephones for a taxi at 2.45 pm, find the probability that she will not be late for the interview.

   b Find, to the nearest minute, the latest time that she should telephone for a taxi in order to have a probability of 0.99 of not being late for the interview.

As well as wishing not to be late, Anita would prefer not to arrive too early, as waiting outside an interview room makes her nervous. The time taken to arrive at her house by a taxi from company *B* may be modelled by a normal distribution with a mean of 12 minutes and a standard deviation of 2 minutes.

c State, giving a reason, which taxi firm you would advise Anita to use.   [AQA(B) Jan 2001]

10 Consultants employed by a large library reported that the time spent in the library by a user could be modelled by a normal distribution with mean 65 minutes and standard deviation 20 minutes.

a Assuming that this model is adequate, what is the probability that a user spends

i less than 90 minutes in the library,

ii between 60 and 90 minutes in the library?

The library closes at 9.00 pm.

b Explain why the model above could not apply to a user who entered the library at 8.00 pm.

c Estimate an approximate latest time of entry for which the model above could still be plausible.
     [AQA(AEB) June 1997]

Please note that this question is NOT from the live examinations for the current specification.

 11 The time taken from entering a self-service canteen at lunchtime to completing the purchase of food may be modelled by a normal distribution with mean 245 seconds and standard deviation 40 seconds.

a Find the probability that it will take Sheila between 200 and 300 seconds from entering the canteen to completing her purchase of food.

b Sheila agrees to purchase her food prior to meeting Kofi in the canteen at 1.00 pm. Find how many seconds before 1.00 pm she should enter the canteen in order to have a probability of 0.98 of meeting Kofi on time.

The service system is reorganised and it is observed that 80% of all customers now complete the purchase of food within 250 seconds of entering the canteen.

c Assuming the standard deviation has remained as 40 seconds, find the new mean time for completing the purchase of food.

d Henry, an impatient customer, claims that the mean time to complete the purchase of food has not changed.

i Assuming that Henry's claim is true, find the new standard deviation.

ii Explain why it is very unlikely that Henry's claim is true.   [AQA(B) June 2001]

5.1 Terminology

Population and sample. Unbiased estimates of a population mean and variance.

In statistics, estimation is the process by which information from a **sample** is used to make statements about a **population**.

One of the conditions for this process is that the sample is **unbiased** (fair) and **representative** (of the population).

For example, in sampling non-fiction books from a library's shelves, the sample should not be restricted to non-fiction paperback books (*biased* as it excludes non-fiction hardback books), nor should the sample be drawn from all the library's books (*unrepresentative* as it could include fiction books).

The most straightforward (unbiased and representative) sampling method, which endeavours to give each member of the population an equal chance of being included in the sample, is called a **simple random sample**.

A simple random sample may be achieved by:

- listing all the N members of the population, called the **sampling frame**
- assigning each member of the population a different number, usually between 0 and $N - 1$ or between 1 and N
- using a table of random numbers to select $n(<N)$ numbers
- discarding repeated numbers and numbers outside the range (0 to $N - 1$ or 1 to N)
- using the n random numbers and the sampling frame to identify the n members of the sample.

In reality, when N is very large, listing all the members is not feasible. A simple random sample is achieved by ensuring that the method for the selection of n members is 'random and representative'.

In many cases, the information from a sample is used to calculate numerical measures such as the mean, mode, median and standard deviation or variance, range and interquartile range. These are known as **(sample) statistics**. Thus, for example, $\bar{x}$ and s^2 are sample statistics.

A characteristic used to define a population is known as a **(population) parameter**. Thus, for example, μ and σ^2 are population parameters.

To estimate the population mean, μ, an obvious choice is the sample mean, $\bar{x}$, so:

- $\bar{x}$ is an estimate of μ
- X is an estimator of μ.

Similarly:

- s^2 is an estimate of σ^2
- S^2 is an estimator of σ^2.

When, in repeated sampling from a population, the average value of a statistic *is equal to* a population parameter, then the estimator is said to be **unbiased**.

Recall:
A **sample** is a set of values selected from a much larger set of values, called a **population**.

Note:
Although you are expected to understand the concept of a simple random sample, this will **not** be tested in the written examination.

Note:
For a particular population, parameters are constants and **not** random variables.

Recall:
An upper case italic letter denotes a random variable, whereas a lower case italic letter denotes a value of the random variable.

In fact:

1 $\bar{x}$ is an unbiased estimate of μ
2 s^2 is an unbiased estimate of σ^2

or

3 $\overline{X}$ is an unbiased estimator of μ
4 S^2 is an unbiased estimator of σ^2.

Note:
This unbiased property is the reason for the divisor $(n-1)$, rather than n, in the formula for the sample variance.

Note:
Statements 3 and 4 are provided in the examination.

5.2 Distribution of the sample mean

The sampling distribution of the mean of a random sample from a normal distribution.
A normal distribution as an approximation to the sampling distribution of the mean of a large sample from any distribution.

If $X \sim N(\mu, \sigma^2)$ then

$$\overline{X} \sim N\left(\mu, \frac{\sigma^2}{n}\right) \quad \text{or} \quad Z = \frac{\overline{X} - \mu}{\frac{\sigma}{\sqrt{n}}} \sim N(0, 1),$$

where $\overline{X}$ denotes the mean of a random sample of size n.

The standard deviation of $\overline{X}$ $\left(\text{i.e. } \frac{\sigma}{\sqrt{n}}\right)$ is sometimes called the **standard error** of the mean.

Note:
The alternative result for Z is provided in the examination.

Example 5.1 The volume of cola in a can may be assumed to be normally distributed with a mean of 336 ml and a standard deviation of 3 ml. For a random sample of six cans:

a write down the distribution for the mean volume of cola per can and find the value for this mean's standard error,
b determine the probability that the mean volume of cola per can exceeds 335 ml.

Step 1: Write down the distribution of X.

Step 2: Write down the distribution of $\overline{X}$ and hence the value of $\frac{\sigma}{\sqrt{n}}$.

Step 3: Write down the form for standardising $\overline{X}$ and hence standardise the given $\bar{x}$-value.

Step 4: Use Table 3 in the examination booklet to find the required probability (checking its 'size').

a Let X = volume of cola in a can, then $X \sim N(336, 3^2)$.
If $\overline{X}$ denotes the mean volume of cola per can in a random sample of six cans, then $\overline{X} \sim N(336, \frac{9}{6})$ or $N(336, 1.5)$.

Thus the standard error of $\overline{X}$ is $\sqrt{1.5}$ or 1.22 (3 s.f.).

b $P(\overline{X} > 335) = P\left(Z > \frac{335 - 336}{\sqrt{1.5}}\right) = P(Z > -0.82) = P(Z < 0.82)$

$= 0.794$ (3 s.f.)

Recall:
In standardising, the denominator is the standard deviation, not the variance.

If $\overline{X}$ denotes the mean of a random sample of size n from **any (non-normal) population** with mean μ and variance σ^2 then, provided n is sufficiently large, an approximation for the distribution of $\overline{X}$ is

$$N\left(\mu, \frac{\sigma^2}{n}\right)$$

This very useful property of $\overline{X}$ is, in part, a result of the **Central Limit Theorem** which states that:

'Irrespective of the distribution of X, for a sufficiently large sample size, n, the distribution of the sample mean, $\overline{X}$, will be approximately normal.'

Recall:
If $X \sim$ normal then $\overline{X} \sim$ normal for all n.

Tip:
You must know and be able to apply the Central Limit Theorem in the examination.

The theorem's complex proof involves 'letting the sample size tend to infinity' but, in fact, $n \geqslant 30$ is deemed sufficiently large for most practical applications.

Additionally, if the population variance, σ^2, is unknown, then an approximation for the distribution of $\overline{X}$ is $N\left(\mu, \dfrac{s^2}{n}\right)$, where s^2 denotes the sample variance, again provided, in practice, that $n \geqslant 30$.

The estimated standard deviation of $\overline{X}$ $\left(\text{i.e. } \dfrac{s}{\sqrt{n}}\right)$ is sometimes called the **estimated standard error** of the mean.

In summary, if $X \sim (\mu, \sigma^2)$:

then $\overline{X} \sim$ approximately $N\left(\mu, \dfrac{\sigma^2}{n}\right)$, provided $n \geqslant 30$ and σ^2 is known,

or $\overline{X} \sim$ approximately $N\left(\mu, \dfrac{s^2}{n}\right)$, provided $n \geqslant 30$ and σ^2 is unknown.

Note:
This practical value of 30 for n is assumed to be sufficient for the applications of these results in the examination.

Example 5.2 The areas of pieces of glass cut by a DIY store have a mean of 0.247 m^2 and a standard deviation of 0.254 m^2.

a Give one practical reason and one statistical reason why the areas of such pieces of glass are unlikely to be normally distributed.

b Estimate the probability that the mean area of a random sample of 40 pieces of glass cut by the DIY store is less than 0.2 m^2. Justify your method.

a It is likely that many pieces of glass cut will have 'small' areas but only a few will have 'large' areas.

For a normal distribution, approximately two thirds of the values lie within $\mu \pm \sigma$, but $\mu - \sigma = 0.247 - 0.254 < 0$ which is impossible!

Step 1: Using the form for standardising $\overline{X}$, standardise the given $\bar{x}$-value.

Step 2: Use Table 3 to find the required probability (checking its 'size').

b As $n = 40$ the Central Limit Theorem can be applied; the distribution of the sample mean is approximately normal.

Thus $P(\text{mean area} < 0.2) = P\left(Z < \dfrac{0.2 - 0.247}{\dfrac{0.254}{\sqrt{40}}}\right)$

$$= P(Z < -1.17) = P(Z > 1.17)$$
$$= 1 - P(Z < 1.17) = 1 - 0.87900$$
$$= 0.121 \text{ (3 s.f.)}$$

Tip:
Sketch a $N(0, 1)$ curve and shade the area required.

SKILLS CHECK **5A: Distribution of the sample mean**

1 The random variable X is normally distributed with mean 61 and standard deviation 4.8.

  a Determine

    i $P(X < 55)$ **ii** $P(56 < X < 66)$

  b Given that $\overline{X}$ denotes the mean of a random sample of size 9, determine

    i $P(\overline{X} > 62.5)$ **ii** $P(61 < \overline{X} < 63)$

2 The volume of red wine in a bottle may be assumed to be normally distributed with mean 72.5 cl and standard deviation 1.2 cl. For the red wine in a random sample of 12 bottles

  a determine the standard error of the sample mean,

  b determine the probability that the mean volume per bottle is less than 72 cl.

3 The operating times, when fully charged, of re-chargeable back-up batteries in domestic central heating systems may be modelled by a normal distribution with mean 740 minutes and standard deviation 60 minutes.

    a A random sample of six of these batteries, all fully charged, is selected. Determine the probability that their average operating time exceeds 12 hours.

    b A random sample of 36 of these batteries, all fully charged, is selected. Determine the probability that their average operating time exceeds 12 hours.

    c State, with a reason, the effect, if any, on each of your answers to **a** and **b**, if the assumption of normality is invalid.

4 The percentage, Y, of a particular chemical in a naturally occurring compound has a mean of 83 and a standard deviation of 21.

    a Give a reason why Y is unlikely to be normally distributed.

    b Given that 50 samples of the compound are selected at random

       i state, with justification, the approximate distribution of the sample mean, $\bar{Y}$;

      ii estimate $P(80 < \bar{Y} < 85)$.

5 The times taken by Year 6 children to read aloud a piece of text have a mean of 4.7 minutes and a standard deviation of 3.2 minutes.

    a Give a reason why these times are unlikely to be normally distributed.

    b A random sample of 64 Year 6 children is selected. Estimate the probability that their average time to read the piece of text is between 4 minutes and 5 minutes.

 6 The weight of crisps in a packet may be assumed to be normally distributed with a mean of 25.8 grams and a standard deviation of 0.35 grams.

    a Bags contain random samples of 6 packets of these crisps. Determine the probability that the mean weight of crisps per packet in a bag exceeds 26 grams.

    b Super-sized bags contain random samples of 24 packets of these crisps.

       i State the distribution of the mean weight of crisps per packet in a super-sized bag.

      ii Hence determine the probability that the total weight of crisps in a super-sized bag is less than 624 grams.

SKILLS CHECK **5A EXTRA is on the CD**

5.3 Confidence intervals for the mean of a normal distribution with known variance

Confidence intervals for the mean of a normal distribution with known variance.
Inferences from confidence intervals.

A **95% (symmetric) confidence interval** for the mean, μ, of a normal population, with known variance σ^2, is given by:

$$\bar{x} \pm 1.96 \times \frac{\sigma}{\sqrt{n}}$$

where $\bar{x}$ denotes the mean of a random sample of size n.

An alternative expression is:

    (sample mean) $\pm 1.96 \times$ (standard error of sample mean).

Note:
The word 'symmetric' is often omitted.

Note:
The formula for a confidence interval is **not** provided in the examination.

Recall:
From Table 4 $P(Z > 1.96) = P(Z < -1.96) = 0.025$, and so $P(-1.96 < Z < 1.96) = 0.95$ or 95%.

- For 90%, 1.96 is replaced by 1.6449
- For 98%, 1.96 is replaced by 2.3263
- For 99%, 1.96 is replaced by 2.5758
- For 99.8%, 1.96 is replaced by 3.0902.

Tip:
Know how to obtain z-values for any percentage confidence level from Table 4 or from your calculator.

Example 5.3 An analysis of the contents of a random sample of 10 jars of a particular variety of apple sauce reveals the following volumes, in millilitres.

$$262 \quad 258 \quad 259 \quad 252 \quad 259 \quad 260 \quad 264 \quad 260 \quad 258 \quad 259$$

The volume of apple sauce in a jar may be assumed to be normally distributed with a standard deviation of 5 millilitres.

a Determine a 95% confidence interval for the mean volume of apple sauce in a jar, giving the limits to one decimal place.

b Determine a 98% confidence interval for the mean volume of apple sauce in a jar, giving the limits to one decimal place.

c Comment on the claim that the mean volume of apple sauce in a jar is 255 ml.

Step 1: Calculate the sample mean.

Step 2: Use Table 4 in the examination booklet to find the values of z for the required confidence levels.

Step 3: Substitute values for $\bar{x}$, z, σ and n into the formula.

Step 4: Substitute values for $\bar{x}$, z, σ and n into the formula.

Step 5: Compare the claimed mean value with the confidence interval(s).

a Sample mean, $\bar{x} = \dfrac{\sum x}{n} = \dfrac{2591}{10} = 259.1$

Thus a 95% confidence interval is given by

$$259.1 \pm 1.96 \times \frac{5}{\sqrt{10}}$$

i.e. 259.1 ± 3.1 or $(256.0, 262.2)$ ml (1 d.p.).

b A 98% confidence interval is given by

$$259.1 \pm 2.3263 \times \frac{5}{\sqrt{10}}$$

i.e. 259.1 ± 3.7 or $(255.4, 262.8)$ ml (1 d.p.).

c Because the confidence interval in **b**, and thus, by implication, in **a**, does not include 255 ml, it suggests (strongly) that the mean volume of apple sauce in a jar is not 255 ml.

Note:
A confidence interval is **not** a probability statement about μ since μ is a population parameter and so is constant. The **correct interpretation** is that, on average, 95% of such intervals include μ.

Note:
The higher the confidence level, the greater is the width (upper limit – lower limit) of a confidence interval.

Note:
On average, only 2% of 98% confidence intervals **do not** include μ.

SKILLS CHECK **5B: Confidence intervals for the mean of a normal distribution with known variance**

1 A population is normally distributed with mean μ and variance 225.

   a A random sample size of 10, drawn from the population, has a mean of 258. Construct

     i a 90% confidence interval for μ;

     ii a 99% confidence interval for μ.

   b Construct a 90% confidence interval for μ, given that a random sample of size 50, drawn from the population, has a mean of 258.

   c Using your answers to **a** and **b**, comment on the effect on a confidence interval of

     i increasing the percentage confidence,

     ii increasing the sample size.

2 The volume of petrol purchased by customers at a supermarket filling station may be assumed to be normally distributed with a standard deviation of 5 litres. Given that a random sample of 20 customers purchased an average of 17.6 litres of petrol, construct a 95% confidence interval for the mean volume of petrol purchased by customers at this filling station.

3 A butcher buys large ham joints from a wholesaler for cooking then slicing and selling as cooked ham. The weights of the joints may be assumed to be normally distributed with a standard deviation of 1.45 kg. A random sample of 10 joints has the following weights in kilograms.

    9.21   10.63   8.27   12.05   7.94   9.72   10.14   11.44   9.56   9.34

  a Construct a 98% confidence interval for the mean weight of ham joints bought by the butcher.

  b The wholesaler claims that, on average, the ham joints weigh more than 11 kg. Comment on this claim.

4 The volume of apple juice in small bottles may be assumed to be normally distributed with a standard deviation of 10 ml. The volume, in millilitres, of apple juice in each of a random sample of 16 bottles is shown below.

    261   248   255   263   246   247   254   257   244   256   261   256   263   258   259   252

  a Construct a 95% confidence interval for the mean volume of apple juice in a bottle.

  b On each bottle is printed 'Contents 250 ml'. Using the given sample and your confidence interval in **a**, comment on this claim.

5 A fishing club's secretary measures, on a random sample of 30 days, the height, in metres, of the river at a certain point. The river's height at this point may be assumed to be normally distributed with a standard deviation of 0.6 metres.

  a Given that the mean of the sample of 30 measurements is 1.67 metres, construct a 98% confidence interval for the mean height of the river at this point.

  b Comment on the claim made by the fishing club's chairman that the mean height of the river at this point is less than 2 metres.

6 A field is planted with quick-growing willow tree cuttings that are subsequently harvested for fuel. The heights of the trees, two years after planting, may be assumed to be normally distributed with a standard deviation of 35 cm. A random sample of 60 trees, measured two years after planting, has a mean of 1.09 metres.

  a Construct a 99% confidence interval for the mean height of the trees, two years after planting.

  b Comment on the claim that the trees will average at least one metre in height, two years after planting.

SKILLS CHECK **5B EXTRA is on the CD**

5.4 Confidence intervals for the mean of a distribution based upon a large sample

Confidence intervals for the mean of a distribution using a normal approximation.
Inferences from confidence intervals.

An approximate **95% confidence interval** for the mean, μ, of any population, with known variance σ^2, is given by

$$\bar{x} \pm 1.96 \times \frac{\sigma}{\sqrt{n}}$$

where $\bar{x}$ denotes the mean of a random sample of size $n \geqslant 30$.

Tip:
These formulae are the same as those in Section 5.3 but you should be aware of the differences in conditions for their application.

An approximate **95% confidence interval** for the mean, μ, of any population, with unknown variance, is given by

$$\bar{x} \pm 1.96 \times \frac{s}{\sqrt{n}}$$

where $\bar{x}$ and s denote the mean and standard deviation, respectively, of a random sample of size $n \geqslant 30$. An alternative expression is:

(sample mean) $\pm$ 1.96 $\times$ (estimated standard error of sample mean).

Example 5.4 The duration of a commuter's car journey to work is known to have a standard deviation of 25 minutes. From a random sample of 65 journeys to work the mean duration is found to be 33.4 minutes.

   a Determine a 90% confidence interval for the mean duration of the commuter's car journeys to work. Justify any assumptions that you make.

   b Comment on the commuter's claim that the mean duration of his car journey to work is longer than 30 minutes.

   c What would be the effect, if any, on your answers if the 65 journeys were recorded each weekday (Monday to Friday) over a 13-week period?

Step 1: As $n = 65 > 30$, we can apply the Central Limit Theorem.

   a The sample mean can be assumed to be (approximately) normally distributed as the sample size is greater than 30; i.e. by Central Limit Theorem.

Step 2: Substitute values for $\bar{x}$, z, σ and n into the formula.

Thus a 90% confidence interval is given by $33.4 \pm 1.6449 \times \dfrac{25}{\sqrt{65}}$

i.e. 33.4 ± 5.1 or $(28.3, 38.5)$ minutes (3 s.f.).

Recall:
The use of Table 4 to obtain 1.6449 for a 90% confidence interval.

Step 3: Compare the claimed mean value with the confidence interval.

   b The commuter's claim that the mean duration of his car journey to work is longer than 30 minutes is unlikely to be valid as the confidence interval includes the value of 30.

Step 4: Indicate the dangers of using non-random samples.

   c The sample would not be random. This would cast doubt on the answers to **a** and hence on the conclusion in **b**.

Example 5.5 A mobile machine saw is set to cut felled tree trunks into logs that are 2.4 metres in length for transport to a sawmill. An analysis of the lengths, x metres, of a random sample of 75 logs cut by the machine gives

$$\sum x = 192.75 \text{ and } \sum (x - \bar{x})^2 = 0.7549$$

   a Calculate values for the mean and standard deviation of the lengths of logs in the sample.

   b Calculate the estimated standard error of the sample mean.

   c Determine a 99% confidence interval for the mean length of logs cut by the machine.

   d What can be deduced from your confidence interval?

   e State the width of your confidence interval.

   f Determine the confidence associated with a width of 0.024 metres.

Step 1: Use the given quantities to find values for $\bar{x}$ and s.

a The mean, $\bar{x} = \dfrac{192.75}{75} = 2.57$

The standard deviation, $s = \sqrt{\dfrac{0.7549}{74}} = 0.101$

Step 2: Calculate $\dfrac{s}{\sqrt{n}}$.

b Estimated standard error of sample mean $= \dfrac{0.101}{\sqrt{75}} = 0.011662$

Step 3: Substitute values for $\bar{x}$, z and $\dfrac{s}{\sqrt{n}}$ into the formula.

c A 99% confidence interval is given by $2.57 \pm 2.5758 \times 0.011662$
i.e. 2.57 ± 0.03 or $(2.54, 2.60)$ metres (3 s.f.).

Step 4: Compare the target mean value with the confidence interval.

d The saw is cutting logs with a mean length greater than 2.4 metres.

Step 5: Calculate the difference between the upper and lower confidence limits.

e The width of the confidence interval is
$2.60 - 2.54$ (or 2×0.03) $= 0.060$ metres

Step 6: Equate the width of confidence interval with z unknown to the required width.

f We require $2 \times \left(z \times \dfrac{0.101}{\sqrt{75}} \right) = 0.024$

Step 7: Solve the equation for z.

Thus $z = \dfrac{0.024 \times \sqrt{75}}{0.101 \times 2} = 1.03$ (2 d.p.)

Step 8: Use Table 1 to find $P(Z < z)$ and hence find $P(-z < Z < z)$.

However, $P(Z < 1.03) = 0.84849$
So $P(Z > 1.03) = 1 - 0.84848 = 0.15151$
Thus $P(-1.03 < Z < 1.03) = 1 - 2 \times 0.15151$
$= 1 - 0.30302$
$= 0.69698$ or 70%

SKILLS CHECK **5C: Confidence intervals for the mean of a distribution based upon a large sample**

1 The volume, x ml, of coffee dispensed by a one-cup coffee pod machine is recorded on a random sample of 48 occasions with the following results, where $\bar{x}$ denotes the sample mean.

$$\sum x = 6576 \qquad \sum(x - \bar{x})^2 = 1692$$

Construct a 95% confidence interval for the mean volume of coffee dispensed by the machine, giving the limits to one decimal place.

2 The weights of a random sample of 50 men, all aged 60 years, have a mean of 82.6 kg and a standard deviation of 26.3 kg.

a Explain why these weights are unlikely to be normally distributed.

b Construct a 90% confidence interval for the mean weight of 60-year-old men.

3 A company, that offers customers a mail order facility, selects a random sample of 100 mail orders. It finds that, for the sample, the average value of an order is £26.35 and the standard deviation is £19.77.

a Explain why this information suggests that the values of mail orders are unlikely to be normally distributed.

b Construct a 98% confidence interval for the average value of the company's mail orders.

c The average value of a customer's purchases in the company's shops is £11.84. Comment on the claim that, on average, the value of a mail order is $2\frac{1}{2}$ times that of a customer's shop purchases.

4 An analysis of a random sample of 50 flights, operated by a particular company, between the Isle of Man and Liverpool shows a mean flight time of 32.8 minutes and a standard deviation of 6.6 minutes.

   a Construct a 99% confidence interval for the company's mean flight time between the Isle of Man and Liverpool.

   b Hence comment on the company's advertising claim that 'Liverpool is less than 35 minutes away'.

 5 Annabel suspects that, on average, it takes her longer to drive from work to home in the evening than it does for her to drive from home to work in the morning. Over a period of time she has established that her time to drive from home to work in the morning is normally distributed with a mean of 28.5 minutes. An analysis of a random sample of 65 of her journeys from work to home in the evening shows an average driving time of 31.6 minutes and a standard deviation of 9.3 minutes.

   a **i** Construct a 90% confidence interval for Annabel's average time to drive from work to home in the evening.

      ii Hence comment on her suspicion.

      iii Why was it not necessary to assume that her driving times from work to home in the evening are normally distributed?

   b State Annabel's median driving time from home to work in the morning.

   c Assuming that Annabel's driving times from work to home in the evening are normally distributed, estimate

      i the 90th percentile   **ii** the interquartile range.

6 As part of a campaign to have the speed limit reduced outside a school from 30 mph to 20 mph, a parent governor recorded the speeds, x mph, of a random sample of 50 vehicles passing the school entrance. Subsequent computations gave:

$$\sum x = 1715 \qquad \sum(x - \bar{x})^2 = 1324.96$$

where $\bar{x}$ denotes the sample mean.

   a Calculate unbiased estimates of the mean and variance of the speed of the traffic passing the school entrance.

   b The speeds of vehicles passing the school entrance may be assumed to be normally distributed.

      i Construct a 99% confidence interval for the mean speed of vehicles passing the school entrance.

      ii Construct an interval within which approximately 99% of the speeds of vehicles passing the school entrance will lie.

      iii Hence comment on the campaign's claim that 'the 30 mph limit is being ignored'.

   c What would be the effect, if any, on your answers in part **b**, if the assumption of normality could not be made?

SKILLS CHECK **5C EXTRA** is on the CD

Examination practice   Estimation

1 The breaking strengths of cables produced by a factory are normally distributed with a mean of 8440 newtons and a standard deviation of 120 newtons. The factory supplies 9 of these cables, selected at random, to a customer.

Determine the probability that the mean breaking strength of the cables supplied is less than 8500 newtons.   [AQA(NEAB) March 2001]

Please note that this question is NOT from the live examinations for the current specification.

2 A firm has cars available for hire. The distance travelled per day by a hire car has a mean of 236 miles and a standard deviation of 80 miles.

a For a random sample of 100 such distances, determine the probability that its mean is less than 250 miles.

b Name the theorem which you have used, and explain why it is applicable in this case.

[AQA(NEAB) March 2000]

Please note that this question is NOT from the live examinations for the current specification.

3 The lengths of steel pins cut by a machine are normally distributed with a mean of 46.2 mm and a standard deviation of 1.5 mm.

a For a sample of four pins cut by the machine, find the probability that their mean length is greater than 45.0 mm.

b State the assumption that you needed to make about the sample when answering part **a**.

[AQA(NEAB) June 1999]

Please note that this question is NOT from the live examinations for the current specification.

4 A machine fills bottles with vinegar such that the volume of vinegar in a bottle has a standard deviation of 8.0 ml. A random sample of 100 bottles of the vinegar is found to contain a mean volume of 356.5 ml per bottle.

Construct a symmetric 90% confidence interval for the mean volume of vinegar in bottles filled by the machine.

[AQA(NEAB) June 2000]

Please note that this question is NOT from the live examinations for the current specification.

5 The volume of soft drink in a can is a normally distributed random variable with a mean of 336 ml and a standard deviation of 3.2 ml.

a A random sample of 16 cans of soft drink is taken. Determine the probability that the mean volume of drink per can exceeds 335 ml.

b State why, in answering part **a**, you did not need to use the Central Limit Theorem.

[AQA(NEAB) June 2001]

Please note that this question is NOT from the live examinations for the current specification.

 6 a Machine *A* produces plastic balls which have diameters that are normally distributed with a mean of 10.3 cm and a standard deviation of 0.16 cm.

   i Determine the proportion of balls which have diameters less than 10.5 cm.

   ii Calculate the diameter exceeded by 75 per cent of balls.

b Machine *B* produces beach balls which have diameters that are normally distributed with a mean of μ cm and a standard deviation of 0.24 cm.

The diameter, *d* cm, of each ball in a random sample of 144 beach balls was measured. This gave:

$$\sum d = 3585.6.$$

   i Calculate an unbiased estimate of μ.

   ii Calculate the standard error of your unbiased estimate.

   iii Construct a 95% confidence interval for μ, giving its limits to two decimal places.

   iv Hence state, with a reason, whether you agree with a claim that $\mu = 25$.

[AQA(A) June 2002]

7 A company has three machines, I, II and III, each producing chocolate bar ice creams of a particular variety.

  a Machine I produces bars whose weights are normally distributed with a mean of 48.1 grams and a standard deviation of 0.25 grams. Determine the probability that the weight of a randomly selected bar is less than 47.5 grams.

  b Machine II produces bars whose weights are normally distributed with a standard deviation of 0.32 grams. Given also that 85 per cent of bars have weights below 50.0 grams, determine the mean weight of bars.

  c From a random sample of 36 bars, selected at random from those produced on Machine III, calculations gave a mean weight of 52.46 grams and an unbiased estimate of the population variance of 0.1764 grams^2.
  i Construct a 95% confidence interval for the mean weight of bars produced on Machine III, giving the limits to two decimal places.
  ii Name the theorem that you have used and explain why it was applicable in this case.

  [AQA(A) June 2001]

 8 The contents of jars of honey may be assumed to be normally distributed with standard deviation 3.1 grams. The contents, in grams, of a random sample of eight jars were as follows:

  458   450   457   456   460   459   458   456.

  a Calculate a 95% confidence interval for the mean contents of all jars.

  b On each jar it states 'Contents 454 grams'. Comment on this statement using the given sample and your results in part **a**.

  c Given that the mean contents of all jars is 454 grams, state the probability that a 95% confidence interval calculated from the contents of a random sample of jars will **not** contain 454 grams.

  [AQA(B) June 2001]

9 A firm is considering providing an unlimited supply of free bottled water for employees to drink during working hours. To estimate how much bottled water is likely to be consumed, a pilot study is undertaken. On a particular day-shift, ten employees are provided with unlimited bottled water. The amount each one consumes is monitored. The amounts, in ml, consumed by these ten employees are as follows:

  110   0   640   790   1120   0   0   2010   830   770

  a Assuming the data may be regarded as a random sample from a normal distribution with standard deviation 510, calculate a 95% confidence interval for the mean amount consumed on a day-shift.

  b i Give a reason, based on the data collected, why the normal distribution may not provide a suitable model for the amount of free bottled water which would be consumed by employees of the firm.
  ii A normal distribution may provide an adequate model but cannot provide an exact model for the amount of bottled water consumed. Explain this statement, giving a reason which does not depend on the data collected.

  c Following the pilot study the firm offers the bottled water to all the 135 employees who work on the night-shift. The amounts they consume on the first night have a mean of 960 ml with a standard deviation of 240 ml.
  i Assuming these data may be regarded as a random sample, calculate a 90% confidence interval for the mean amount consumed on a night-shift.
  ii Explain why it was **not** necessary to know that the data came from a normal distribution in order to calculate the confidence interval in part **c i**.
  iii Give **one** reason why it may be unrealistic to regard the data as a random sample of the amounts that would be consumed by all employees if the scheme was introduced on all shifts on a permanent basis.

  [AQA(B) Jan 2002]

Correlation and regression

6.1 Product moment correlation coefficient

Calculation and interpretation of the product moment correlation coefficient. Linear scaling.

The term **correlation** is used to imply that there is a **linear relationship** between two random variables, X and Y. The strength of this linear relationship may be seen from a **scatter diagram**. The following seven diagrams provide illustrations.

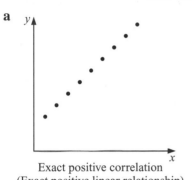

a

Exact positive correlation
(Exact positive linear relationship)

b

Strong positive correlation
(Clear positive linear relationship)

c

Little or no correlation
(No linear relationship)

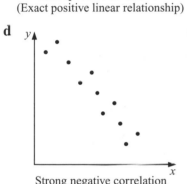

d

Strong negative correlation
(Clear negative linear relationship)

e

Exact negative correlation
(Exact negative linear relationship)

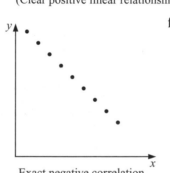

f

Non-linear relationship

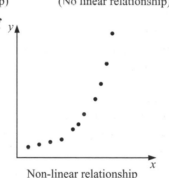

g

Non-linear relationship

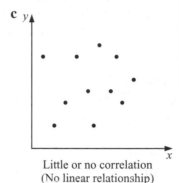
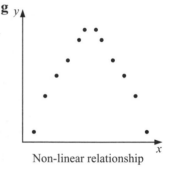

Calculating a value of the product moment correlation coefficient

The **product moment correlation coefficient**, denoted by r, is used to measure the strength of linear correlation between two random variables, X and Y.

For n pairs of data points (x_i, y_i), r is defined by:

$$r = \frac{\sum(x - \bar{x})(y - \bar{y})}{\sqrt{\{\sum(x - \bar{x})^2\}\{\sum(y - \bar{y})^2\}}} \quad \text{or} \quad r = \frac{S_{xy}}{\sqrt{S_{xx} \times S_{yy}}} \quad \text{where}$$

$$S_{xx} = \sum(x - \bar{x})^2 = \sum x^2 - \frac{(\sum x)^2}{n}$$

$$S_{yy} = \sum(y - \bar{y})^2 = \sum y^2 - \frac{(\sum y)^2}{n} \quad \text{and}$$

$$S_{xy} = \sum(x - \bar{x})(y - \bar{y}) = \sum xy - \frac{(\sum x)(\sum y)}{n}$$

Combining the definitions above, a commonly used formula for calculating r is given by:

$$r = \frac{\sum xy - \dfrac{(\sum x)(\sum y)}{n}}{\sqrt{\left(\sum x^2 - \dfrac{(\sum x)^2}{n}\right)\left(\sum y^2 - \dfrac{(\sum y)^2}{n}\right)}}$$

Recall:

The definitions of $\bar{x}$ and $\bar{y}$.

Note:

This formula is provided in the examination.

Note:

Scientific and graphical calculators may be used in the examination and should have a correlation coefficient function (r). Use of this function in the examination is *encouraged*.

Example 6.1 Students on a Spanish course were given a listening test and a written examination. For the listening test students were given a mark out of 25, and for the written examination they were given a mark out of 75. The results for a random sample of eight students are shown in the table.

Student	1	2	3	4	5	6	7	8
Listening test mark	10	21	22	18	17	14	20	16
Written examination mark	21	53	35	45	51	33	59	40

Calculate the value of the correlation coefficient between the results in the listening test and written examination.

Step 1: Label the marks as x and y.

Let x denote a mark in the listening test and let y denote a mark in the written examination.

Step 2: Construct and complete a table to give the required totals.

Student	x	y	x^2	y^2	xy
1	10	21	100	441	210
2	21	53	441	2809	1113
3	22	35	484	1225	770
4	18	45	324	2025	810
5	17	51	289	2601	867
6	14	33	196	1089	462
7	20	59	400	3481	1180
8	16	40	256	1600	640
Total	138	337	2490	15 271	6052

Step 3: Substitute the totals into the computational formula.

Thus:

$$r = \frac{6052 - \dfrac{138 \times 337}{8}}{\sqrt{\left(2490 - \dfrac{138^2}{8}\right)\left(15\,271 - \dfrac{337^2}{8}\right)}}$$

$$= \frac{238.75}{\sqrt{109.5 \times 1074.875}} = 0.696 \text{ (3 s.f.)}$$

Tip:

Get to know the correlation coefficient function on your calculator by checking this answer.

6A: Calculating a value of the product moment correlation coefficient

1 The length and maximum diameter of each of a sample of six courgettes, selected at random from a box, are measured, in millimetres, with the following results.

Length	167	185	197	155	178	195
Diameter	36	38	34	29	37	36

Calculate the value of the product moment correlation coefficient between the length and diameter of these courgettes.

2 The dominant hand span, x mm, and dominant foot length, y mm, of each of a random sample of 10 adult males are tabled below.

x	203	206	234	207	218	222	265	248	231	225
y	252	257	271	255	259	263	267	265	262	259

Calculate the value of the product moment correlation coefficient for these data.

 3 The accounts of a jobbing builder show, for each of a random sample of 10 jobs, the weights, in kilograms, of sand and cement charged.

Sand	125	200	350	175	300	150	325	225	250	100
Cement	25.0	50.0	75.0	37.5	62.5	37.5	75.0	37.5	62.5	25.0

Calculate the value of the product moment correlation coefficient for these data.

4 Nine leylandii trees, of the same age and approximate height, were planted to form a hedge. Four years later, their heights, x metres, and maximum trunk diameters, y mm, were measured with the following results.

x	1.67	1.58	1.36	1.84	1.33	1.02	1.78	1.55	1.64
y	32	47	42	28	35	39	42	33	44

Calculate the value of the product moment correlation coefficient between the height and maximum trunk diameter of these nine leylandii trees.

 5 A householder records, on each of a random sample of 12 days during winter, the outside air temperature at 0600 and 1800 together with the volume of gas used by the property's central heating boiler from 0000 to 2400. The table shows, for each selected day, the average of the two temperatures, $x\,°C$, and the gas used, y kWh.

x	5.5	3.0	7.0	4.0	3.5	1.0	−0.5	2.0	5.0	4.5	6.0	7.0
y	64	65	45	87	83	102	115	94	73	75	54	49

Calculate the value of the product moment correlation coefficient for these data.

6 At a randomly chosen hour between 0700 and 2100 on each of a random sample of 10 weekdays, a note is made at a main line station of the differences, in minutes, between the scheduled and actual arrival times of the next northbound inter-city train and of the next southbound inter-city train. The results are tabled below, with negative values indicating late arrivals.

Northbound	0.0	2.1	1.5	0.5	−5.6	4.2	−5.2	1.6	0.7	−2.8
Southbound	−2.4	−6.3	3.5	2.3	0.3	2.3	−6.8	−5.7	1.5	2.9

For these data, calculate the value of the product moment correlation coefficient.

SKILLS CHECK **6A EXTRA** is on the CD

Interpreting a value of the product moment correlation coefficient

It is always the case that $-1 \leqslant r \leqslant +1$.

$r = +1$ indicates that **all the points** in the scatter diagram **lie exactly on a straight line** having a **positive slope**, as illustrated in the scatter diagram labelled **a** on page 67

$r = -1$ indicates that **all the points** in the scatter diagram **lie exactly on a straight line** having a **negative slope**, as illustrated in the scatter diagram labelled **e** on page 67

$r = 0$ indicates that the points on the scatter diagram reveal that there is **no linear relationship** between the pairs of values, as illustrated in the scatter diagrams labelled **c** and **g** on page 67.

> **Note:**
> $r = 0$ does not imply **no relationship** but only **no linear** relationship.

- For the scatter diagram labelled **b**, $r \approx +0.95$ indicating strong positive correlation.

- For the scatter diagram labelled **d**, $r \approx -0.95$ indicating strong negative correlation.

- For the scatter diagram labelled **f**, $r \approx +0.90$ but the relationship is clearly non-linear. This illustrates one of the reasons why care is needed in interpreting a value of r, particularly without reference to the corresponding scatter diagram.

> **Note:**
> A 'large' value of r may result from a non-linear relationship.

Some other values of r are illustrated in the following three scatter diagrams.

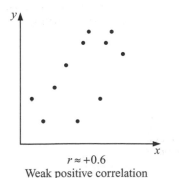

$r \approx +0.6$
Weak positive correlation
(Little evidence of a positive linear relationship)

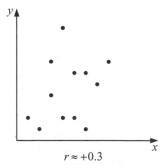

$r \approx +0.3$
Weak positive correlation
(No real evidence of a linear relationship)

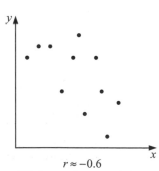

$r \approx -0.6$
Weak negative correlation
(Little evidence of a negative linear relationship)

Example 6.2 As part of a comparative study into the handling of two different models of car, Model I and Model II, nine drivers were each asked independently to 'parallel-park' each model in the same space. Their times, in seconds, are given in the table.

Driver	A	B	C	D	E	F	G	H	I
Time to park Model I (x)	36	30	39	22	34	26	47	23	41
Time to park Model II (y)	32	35	42	34	41	30	42	27	43

a Calculate the value of the product moment correlation coefficient between the two sets of times.

b Interpret, in context, the value obtained in **a**.

c Plot a scatter diagram of these data.

d Explain how the scatter diagram in **c** provides confirmation of the value obtained in **a**.

Step 1: Calculate r (using the function on your calculator).

a Using a calculator, with a correlation function, gives

$$r = +0.800 \text{ (3 s.f.)}$$

Tip:

Use your calculator to check this answer.

Step 2: Interpret the value obtained.

b This suggests that drivers who park Model I cars quickly, often also tend to park Model II cars quickly, and vice versa.

Step 3: Plot a scatter diagram.

c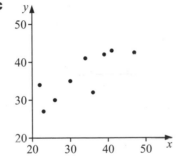

Note

In scatter diagrams, the scales need not start at 0.

Tip:

Be aware of what a particular value of r suggests about the corresponding diagram, and vice versa.

Step 4: Compare the calculated value of r with the value indicated by the scatter diagram.

d The scatter diagram reveals a fairly high level of positive linear correlation and this is consistent with $r = +0.800$.

Limitations of the product moment correlation coefficient

Unusual or freak values within the data, called **outliers**, can drastically alter the value of r as illustrated in the following two scatter diagram.

Recall:

r only measures the strength of a **linear** relationship.

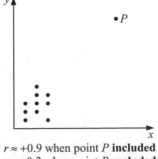

$r \approx +0.9$ when point P **included**
$r \approx +0.2$ when point P **excluded**

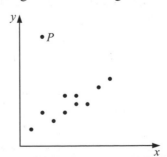

$r \approx +0.2$ when point P **included**
$r \approx +0.9$ when point P **excluded**

Such values, often identifiable from a scatter diagram, should be investigated and reasons sought for their inclusion or exclusion.

It is possible to find that two variables appear to be highly correlated and yet there is no sensible reason why this is the case. In other words, there is **no cause and effect** and in such cases the correlation is termed '**spurious**'. In many instances, the effect of a third variable, such as time, is the cause of the spurious correlation.

Example 6.3 For his project, Walter, a social sciences student, obtains information on a random sample of children at a local primary school. From the sample he calculates the value of the correlation coefficient:

a between a child's age and a child's height to be -0.437

b between a child's height and a child's weight to be 0.763

c between a child's height and a child's average hand-span to be 1.142

d between a child's height and reading ability score to be 0.621.

Classify each of these statements as plausible, spurious, probably incorrect or definitely incorrect. Give a reason in each case.

Step 1: Check whether $-1 \leqslant r \leqslant 1$.

Step 2: If 'no', then definitely incorrect.

Step 3: If 'yes', consider the likely (range of) values of r from the context.

Step 4: Compare this with the given value.

a Probably incorrect. In general, height increases with age for primary schoolchildren.

b Plausible. In general, the taller a child, the greater the child's weight.

c Definitely incorrect. Value cannot be greater than 1.

d Spurious. Most likely that both variables increase with a child's age.

Tip:

Be aware of what a particular value of r suggests about the relationship.

Linear scaling and the product moment correlation coefficient

Linear scaling of either x or y or both has no effect on the value of the product moment correlation coefficient.

Example 6.4 The height, x metres, and diameter, y centimetres, of each of a garden nursery's particular variety of cut christmas tree was measured. The results were as follows.

Mean height $= 1.7$ metres
Standard deviation of height $= 0.2$ metres

Mean diameter $= 55$ centimetres
Standard deviation of diameter $= 5$ centimetres

Product moment correlation coefficient between height and diameter $= 0.655$

If both height and diameter had been measured in centimetres, write down the value of:

a the mean height

b the standard deviation of height

c the product moment correlation coefficient between height and diameter.

Step 1: Multiply the mean by 100.

a 170

Step 2: Multiple the standard deviation by 100.

b 20

Step 3: No change in the value of r.

c 0.655

Recall:

Linear scaling of numerical measures in Section 1.2.

1 Estimate, without undertaking any calculations, the value of the product moment correlation coefficient between the variables x and y in each of the following scatter diagrams.

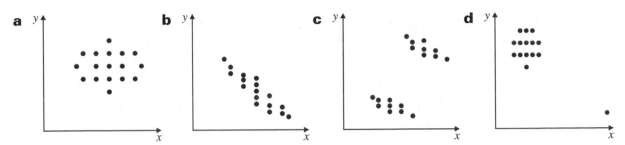

2 In the machine sewing section of a high fashion clothes company, a score is assigned to each finished item on the basis of its quality (the better the quality, the higher the score). Each seamstress's pay is partly dependent on the number of items she completes. The number, x, of items completed by 10 seamstresses during a particular week and their mean quality scores are shown below.

Seamstress	A	B	C	D	E	F	G	H	I	J
x	14	13	17	16	18	12	32	18	15	19
y	7.8	8.3	7.7	6.8	8.8	6.3	3.7	7.5	8.6	6.8

a Calculate the value of the product moment correlation coefficient for these data.

b Interpret, in context, your value.

c i Plot the data on a scatter diagram.

   ii Hence discuss, briefly, whether or not your interpretation in **b** should now be amended.

d When the results were presented at a Quality Assurance meeting, the Personnel Officer explained that seamstress G was experiencing severe personal financial difficulties. Explain the implication of this additional information on your conclusions.

3 As part of an investigation into a possible association between characteristics of the human body, the following data on the chest measurement, x cm, and waist measurement, y cm, of each of 12 individuals were collected.

Individual	A	B	C	D	E	F	G	H	I	J	K	L
x	102	98	111	93	98	103	101	96	95	96	108	90
y	95	86	95	65	70	89	91	62	63	68	93	59

a i Calculate the value of the product moment correlation coefficient for these data.

   ii Interpret, in context, your value.

b Plot a scatter diagram of the data.

c In fact, half of the 12 individuals are male and half are female.

   i With reference to your scatter diagram, write down the letters referencing the individuals that you consider to be **female**.

   ii Give a reason why you have chosen this set of individuals as females.

d Suggest a next possible step in the analysis.

4 In a comparison of two methods for measuring the consumption rates of diesel engines, a random sample of eight engines was used. For each engine, the consumption rate, in litres per hour, was measured by an established method, Method A, and by an experimental method, Method B, with the following results.

Engine	1	2	3	4	5	6	7	8
Method A	21	113	57	53	156	75	100	33
Method B	42	148	84	76	161	91	102	48

    a Plot a scatter diagram of these data.

    b Calculate the value of the product moment correlation coefficient, r, for these data.

    c Explain how your scatter diagram confirms your value of r.

    d Explain why, despite your value of r, it may be claimed that Method B does not measure consumption rates accurately.

5 A psychologist, working with teenagers in a remand centre, was interested in studying the possible relationship between intelligence and delinquency. A delinquency index, DI, (ranging from 0 to 50) was formulated to measure both the severity and frequency of crimes committed, whilst intelligence was measured by IQ. The table below shows the values of IQ and DI for a random sample of 11 teenagers.

Teenager	A	B	C	D	E	F	G	H	I	J	K
IQ	89	100	121	73	79	109	134	84	95	114	80
DI	32	26	22	38	39	25	45	36	32	23	32

    a **i** Calculate the value of the product moment correlation coefficient for these data.

       ii Interpret, in context, your value.

    b **i** Plot a scatter diagram of the data.

       ii State whether or not your interpretation in **a ii** should be amended.

    c It was subsequently discovered that the value of DI for teenager G was 15 and not 45 as previously recorded. Without further calculations:

       i explain the implications of this additional information on your conclusions,

       ii estimate the value of the product moment correlation coefficient between IQ and DI for teenagers in remand centres.

6 **a** Sketch a scatter diagram that might represent the results that gave a value for the product moment correlation coefficient of:

       i -0.85,   **ii** 0.06.

    b Assuming no errors in calculation, why might each of the following statements, regarding a value of the product moment correlation coefficient, be incorrect?

       i 'A correlation of -0.0735 indicates that the two variables are not related.'

       ii 'A correlation of 0.927 indicates that an increase in the value of one variable causes an increase in the value of the other variable.'

7 Classify each of the following statements, regarding a value of the product moment correlation coefficient, as definitely incorrect, probably incorrect, plausible or spurious. Justify each of your choices.

    a Between the tail length and the weight of female dormice it is 0.413.

    b Between a household's quarterly consumptions of gas and electricity it is -1.129.

    c Between the diameters and weights of yellow melons it is -0.642.

    d Between a dentist's weekly sales of toothpaste and weekly number of patients' fillings it is 0.428.

    e Between the maximum daily temperature and a café's daily sales of hot drinks it is -0.427.

SKILLS CHECK **6B EXTRA** is on the CD

Identification of response (dependent) and explanatory (independent) variables in regression.
Calculation of least squares regression lines with one explanatory variable.
Scatter diagrams and drawing a regression line thereon.
Calculation of residuals. Linear scaling.

Here, the term **regression** is used to imply the determination of the actual equation for the linear relationship or **straight line** between two variables, X and Y.

The general form for the equation of a straight line for y on x is:

$$y = a + bx$$

where b denotes the **gradient** or **slope** and a denotes the **intercept** with the y-axis.

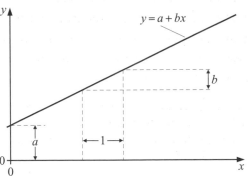

Recall:

Correlation deals with measuring the strength of the linear relationship between two random variables.

Recall:

The form $y = mx + c$.

Identification of response and explanatory variables

In regression, it is important to identify the variables X and Y correctly.

The **response** or **dependent** variable, denoted by Y, is the variable that is likely to be affected by set changes in the **explanatory** or **independent** variable, denoted by X.

Note:

In correlation, interchanging X and Y leaves r unchanged.

Calculating the equation of the least squares regression line

For n pairs of data points (x_i, y_i), the **method of least squares** requires that the values of a and b for the straight line $y = a + bx$, are such that the sum of squares of the **vertical distances of the data points from the line is a minimum**. These vertical distances, called **residuals**, are denoted by r_1, $r_2, r_3, \ldots, r_n$ where $r_i = y_i - (a + bx_i) = y_i - a - bx_i$.

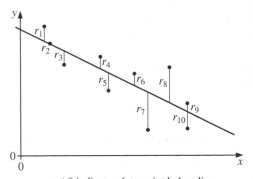

$r_i < 0$ indicates data point below line
$r_i > 0$ indicates data point above line
$r_i = 0$ indicates data point on line

Note:

$\sum r_i^2$ is minimised.

Note:

Do not confuse the r_i, with the correlation coefficient, r.

The values of a and b are given by:

$$b = \frac{\sum(x - \bar{x})(y - \bar{y})}{\sum(x - \bar{x})^2} \quad \text{or} \quad b = \frac{S_{xy}}{S_{xx}} \quad \text{and} \quad a = \bar{y} - b\bar{x} \quad \text{where}$$

$$S_{xx} = \sum(x - \bar{x})^2 = \sum x^2 - \frac{(\sum x)^2}{n} \quad \text{and}$$

$$S_{xy} = \sum(x - \bar{x})(y - \bar{y}) = \sum xy - \frac{(\sum x)(\sum y)}{n}$$

Note:

Most of these formulae are provided in the examination.

Recall:

$\sum x^2 \neq (\sum x)^2$.

Note:

S_{xx} is a sum of squares so **cannot** be negative.

Note:

S_{xy} **can be either** positive or negative.

Example 6.5 As part of an investigation into the durability of car tyres made from a particular rubber compound, simulated wear tests on the tyres were undertaken in a laboratory. The table shows the percentage wear after various set simulated distances, measured in thousands of kilometres.

Distance	15	20	25	30	35	40	45
Wear	26	39	43	55	58	65	79

Calculate the equation of the appropriate least squares regression line.

Step 1: Identify x and y. Here it is wear that is affected by distance; not the reverse! Thus distance is the explanatory variable, x, and wear is the response variable, y.

Note:

Scientific and graphical calculators may be used in the examination and may have regression functions (a and b). Use of these functions in the examination is *encouraged*.

x	y	x^2	xy
15	26	225	390
20	39	400	780
25	43	625	1075
30	55	900	1650
35	58	1225	2030
40	65	1600	2600
45	79	2025	3555
Total	210	365	7000	12080

Step 2: Construct and complete a table to give the required totals.

Warning

Some calculators denote the gradient by a and the intercept by b ($y = ax + b$). Get to know your calculator's notation for regression.

Thus

Step 3: Substitute the totals into the computational formulae.

$$S_{xx} = 7000 - \frac{(210)^2}{7} = 700 \text{ and } S_{xy} = 12\,080 - \frac{210 \times 365}{7} = 1130$$

Hence

Step 4: Calculate values for b and thus a.

$$b = \frac{1130}{700} = 1.61 \text{ (3 s.f.)}$$

$$a = \left(\frac{365}{7}\right) - \left(\frac{1130}{700}\right)\left(\frac{210}{7}\right) = 3.71 \text{ (3 s.f.)}$$

Tip:

Do not round $\bar{y}$, b or $\bar{x}$ to 3 s.f. before substituting into the formula for a.

The least squares regression line for wear (y) on distance (x) is thus given by

$$y = 3.71 + 1.61x$$

Tip:

Get to know the regression functions on your calculator by checking this answer.

SKILLS CHECK **6C: Calculating the equation of the least squares regression line**

1 Calculate the equation of the least squares regression line of y on x for each of the following data sets.

a

x	1	2	3	4	5	6	7	8	9
y	5	7	7	7	9	9	11	13	13

b

x	1	2	3	4	5	6	7	8	9
y	48	46	40	34	32	26	24	22	16

2 To investigate a process that is carried out repeatedly in a chemical works, the amount, x grams, of a chemical added to a mixture is varied and the concentration, y per cent, of the final product is noted. The results are as follows.

x	5	5	10	10	15	15	20	20	25	25
y	2.6	3.0	4.4	4.1	6.4	6.1	8.1	7.3	9.5	10.4

Calculate the equation of the least squares regression line of y on x.

 3 The experimental data below were obtained by measuring the horizontal distance, d cm, rolled by an object released from a fixed point on a plane inclined at an angle, $\theta°$, to the horizontal.

θ	8	26	34	18	20	10	30	16	24	14
d	46	132	170	95	104	53	146	86	112	76

Calculate the equation of the least squares regression line of distance on angle.

4 For a period of three years, a company records, each quarter, the planned number of units of output, x thousand, and the total cost, y, in £000s after taking account of inflation, of their production. The table below shows the results.

x	15	30	55	75	50	25	50	70	20	40	60	80
y	35.2	54.6	76.9	94.8	62.6	44.0	67.2	92.7	38.7	54.6	84.1	104.6

Calculate the equation of the least squares regression line of y on x.

 5 In the development of a new plastic, a variable of interest was its deflection when subjected to a constant force underwater. It was expected that, for a limited range of temperatures, the deflection would be approximately linearly related to the temperature of the water. Thus the deflection was measured at each of a set of predetermined temperatures with the following results.

Temperature	15.0	17.5	20.0	22.5	25.0	27.5	30.0	32.5	35.0	37.5	40.0
Deflection	2.05	2.11	2.39	2.54	2.78	2.91	3.28	3.33	3.56	3.46	3.82

Calculate the equation of the appropriate least squares regression line.

6 An object of weight x grams is attached to one end of a spring. The spring is then hung vertically from its other end and the length of the spring, y millimetres, is recorded. The table shows the results for 9 different weights.

Weight	10	15	20	25	30	35	40	45	50
Length	99.5	102.5	104.2	108.5	111.2	112.9	116.2	120.4	123.6

Calculate the equation of the least squares regression line of length on weight.

SKILLS CHECK **6C EXTRA is on the CD**

Drawing a regression line on a scatter diagram

This is done by:
- choosing two (easy-to-use) values of x, one towards each end of the range of x-values given
- using the regression equation to calculate the corresponding y-values
- plotting these two points on the scatter diagram
- joining these two points by a straight line.

Tip:

If your plotted line does not pass through the point $(\overline{x}, \overline{y})$ **and** does not follow the trend of the points (x, y), **then you have made an error!**

Predictions and extrapolations

The regression equation can be used to:
- **predict** further values of y for values of x **within** the given range of x-values
- **extrapolate** further values of y for values of x **outside** the given range of x-values.

The accuracy of a predicted value depends, in part, on how close the given values are to the fitted line.

The accuracy (even validity) of an extrapolated value depends, in part, on how close the given values are to the fitted line and how far outside the given range of x-values is the extrapolation.

Note:

In the examination, the word 'predict' may infer either prediction or extrapolation.

Note:

The regression functions on your calculator should calculate predictions.

Tip:

Great care should be taken in using extrapolation as it can often lead to a y-value that does not make sense!

Example 6.6 As part of a study into the ageing of photographic film, a sensitometry laboratory collected the following information on the change in blue balance, y, of a particular make of film at increasing periods, x weeks, after manufacture.

x	2	4	6	8	12	16	20	24
y	9.6	11.3	18.7	22.1	32.6	36.2	44.5	49.8

a Plot a scatter diagram of these data.

b Calculate the equation of the least squares regression line of y on x.

c Draw your line on your scatter diagram.

d Predict the change in blue balance of this make of film:
  i 10 weeks after manufacture;
  ii 40 weeks after manufacture.

In each case, comment on the likely accuracy of your estimate.

Step 1: Plot a scatter diagram.

a
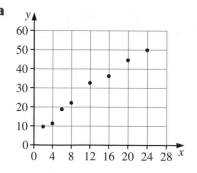

Step 2: Calculate a and b using the functions on your calculator.

b Using a calculator, with regression functions, gives $b = 1.88$ (3 s.f.) and $a = 6.47$ (3 s.f.) or, in other words, $y = 6.47 + 1.88x$.

Step 3: Calculate the coordinates of two points satisfying the equation.

Step 4: Plot the two points on the scatter diagram and join by a straight line.

c Using this equation, when $x = 2$, $y = 10.2$ (3 s.f.) and when $x = 20$, $y = 44.1$ (3 s.f.).

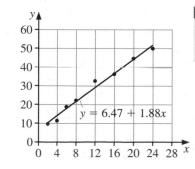

Tip:

Use your calculator to check the answers to **b**, **c** and **d**.

d Using the equation:

    i $y_{10} = 6.47 + 1.88 \times 10 = 25.3$

    ii $y_{40} = 6.47 + 1.88 \times 40 = 81.7$

Note:

y_{10} could be read from the scatter diagram.

Prediction **i** is **likely** to be accurate as 10 lies within the given range of 2 to 24 **and**, from the scatter diagram, the spread of the points about the fitted line is 'small'.

Prediction **ii** involves extrapolation and is **most unlikely** to be accurate as 40 lies well outside the given range of 2 to 24 even though, from the scatter diagram, the spread of the point about the fitted line is 'small'.

Calculation and examination of residuals

The **residual**, denoted by r_i, corresponding to the data point (x_i, y_i) is defined as the vertical distance of the data point from the fitted regression line, $y = a + bx$.

i.e. $r_i = y_i - (a + bx_i) = y_i - a - bx_i$.

- $r_i < 0$ indicates that the data point is **below** the fitted regression line
- $r_i > 0$ indicates that the data point is **above** the fitted regression line
- $r_i = 0$ indicates that the data point is **on** the fitted regression line.

Thus a 'relatively small' residual indicates that the corresponding data point is 'near to' to the fitted line, whereas a 'relatively large' residual indicates that the corresponding data point is 'far away' from the fitted line.

An examination of the residuals, often in conjunction with the scatter diagram, can be used to:

- check the plausibility of the linear model
- identify values with relatively large residuals, called **outliers**, which may have an undue influence on the actual equation of the line
- modify predictions and conclusions in the light of additional information.

Recall:

The method of least squares minimises $\sum r_i^2$.

Note:

$\sum r_i$ is **always zero**.

Note:

The formula for r_i is **not** provided in the examination.

Recall:

Non-linear relationships and the identification and effect of outliers in correlation.

Example 6.7 Tom (T), Dick (D) and Harry (H) are employed by Smiths Car Sales Limited. Each is responsible for cleaning and washing cars at one of the company's three showrooms. The number of cars to be cleaned and washed at each showroom varies from day to day.

The table shows 12 observations of the number, *x*, of cars to be cleaned and washed and the time taken, *y* minutes. The employee carrying out the task is also shown.

Employee	D	H	H	T	D	T	T	D	H	D	H	T
x	20	13	17	15	22	11	24	16	20	17	19	12
y	226	113	148	158	242	110	231	186	162	190	166	131

a Plot a scatter diagram of these data and identify the employee by labelling each point.

b Calculate the equation of the least squares regression line of y on x in the form $y = a + bx$ and draw your line on your scatter diagram.

c Give a practical interpretation, where possible, of your values for a and b.

d Use your equation to estimate the time that would be taken to clean and wash 18 cars.

e Calculate the residuals for the four observations when Harry did the cleaning and washing.

f Modify your answer to **d**, given that Harry is to clean and wash the 18 cars.

Step 1: Plot a scatter diagram and label points. **a**

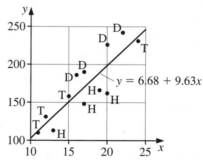

Recall:

In scatter diagrams, the scales need not start at 0.

Step 2: Calculate a and b using the functions on your calculator.

Step 3: Calculate the coordinates of two points satisfying the equation.

Step 4: Plot the two points on the scatter diagram and join them by a straight line.

b Using a calculator, with regression functions, gives $b = 9.63$ (3 s.f.) and $a = 6.68$ (3 s.f.) or, in other words, $y = 6.68 + 9.63x$. Using this equation, when $x = 10$, $y = 106$ (3 s.f.) and when $x = 25$, $y = 247$ (3 s.f.).

$y = 6.68 + 9.63x$

Tip:

Check that your plotted line passes through the point $(\bar{x}, \bar{y})$ **and** follows the trend of the points.

Step 5: Interpret, if sensible, the value of the intercept, a. **c** The value of 6.68 for a has no practical interpretation as it suggests a time of approximately 7 minutes to clean and wash no cars!

Recall:

Extrapolation as $x = 0$ is well outside the range of 11 to 24.

Step 6: Interpret the value of the slope, b. The value of 9.63 for b suggests that, within the observed range for x (11 to 24), the average time taken to clean and wash each (additional) car is approximately 10 minutes.

Step 7: Use the equation to predict a y-value for the given x-value. **d** Using the equation $y_{18} = 6.68 + 9.63 \times 18 = 180$ minutes (3 s.f.) or 3 hours.

Step 8: Calculate the four residuals. **e** $r_2 = 113 - (6.68 + 9.63 \times 13) = -18.9$ (3 s.f.).
$r_3 = 148 - (6.68 + 9.63 \times 17) = -22.4$ (3 s.f.).
$r_9 = 162 - (6.68 + 9.63 \times 20) = -37.3$ (3 s.f.).
$r_{11} = 166 - (6.68 + 9.63 \times 19) = -23.6$ (3 s.f.).

Note:

Keeping more than 3 s.f. for a and b results in minor changes in the residuals.

Step 9: Adjust the estimated time using the average residual. **f** The average of the above four residual values is approximately -26 minutes. Thus the estimated time for Harry to clean and wash 18 cars is $180 - 26 = 154$ minutes.

Linear scaling

The effect of linear scaling in regression is best illustrated by an example.

Recall:
Linear scaling has no effect in correlation.

Example 6.8 The regression equation relating the fuel consumption, y miles per gallon, of a car to its speed, x miles per hour, is given by

$$y = 75.0 - 0.700x \quad \text{for} \quad 40 \leqslant x \leqslant 60$$

Given that there are 1.6093 kilometres in 1 mile, and that 1 mile per gallon is equal to 0.35400 kilometres per litre, find the equivalent equation relating fuel consumption, v kilometres per litre, to the car's speed, u kilometres per hour.

Step 1: Find x in terms of u, and y in terms of v.

Here $u = 1.6093x$ and $v = 0.345y$

so $\quad x = \dfrac{u}{1.6093}$ and $y = \dfrac{v}{0.354}$

Step 2: Substitute for x and y in the regression equation, then simplify.

Thus $y = 75.0 - 0.700x$ becomes:

$$\frac{v}{0.354} = 75.0 - 0.700 \times \left(\frac{u}{1.6093}\right) \text{ or } v = 26.6 - 0.154u \text{ (3 s.f.)}$$

Tip:
Check the final simplification using your calculator.

SKILLS CHECK **6D: Regression analysis**

1 In an attempt to increase the yield, y kg/h, of a chemical process, a technician varies the percentage, x, of a certain additive whilst keeping other conditions as constant as possible. The results are tabled below.

x	2.0	2.5	3.0	3.5	4.0	4.5	5.0	5.5	6.0
y	123	125	134	130	137	134	141	139	144

a Plot a scatter diagram of these data.

b i Calculate the equation of the least squares regression line in the form $y = a + bx$.

   ii Draw the line on your scatter diagram.

c Interpret, in context, your values for a and b.

d Estimate the yield when the percentage of additive is

   i 4.25,   **ii** 15.

In each case, comment on the likely accuracy of your estimate.

2 The following data were collected during a study, under experimental conditions, of the effect of temperature, $x\,°\text{C}$, on the pH, y, of skimmed milk.

x	5	10	15	20	25	30	35	40	45	50
y	6.85	6.82	6.78	6.67	6.63	6.65	6.61	6.52	6.54	6.48

a Plot a scatter diagram of these data.

b i Calculate the equation of the least squares regression line in the form $y = a + bx$.

   ii Draw the line on your scatter diagram.

c Interpret, in context, your values for a and b.

d Estimate the pH of skimmed milk when the temperature is:

   i 37.5 °C;   **ii** 75 °C.

In each case, comment on the likely accuracy of your estimate.

3 At a certain location, the sand content, in per cent, of the soil was measured at each of eight accurately measured depths, in centimetres, with the tabled results below.

Depth	10	15	20	30	40	50	75	100
Sand content	94	90	85	72	66	55	20	6

a Identify, with a reason, the explanatory variable, x, and the response variable, y.

b Plot a scatter diagram of these data.

c **i** Calculate the equation of the least squares regression line in the form $y = a + bx$.
ii Draw the line on your scatter diagram.

d Interpret, in context, your value for b.

e Use your regression line to estimate the sand content of the soil at the surface. Comment on the value you obtain.

4 Sets of crockery are packed individually to meet customers' requirements. The packaging manager introduced a new procedure in which each packer was responsible for all stages of an order from its receipt to dispatch. In order to be able to estimate the time needed to deal with particular orders, she recorded the time taken, y minutes, by an experienced packer to deal with his first 12 orders using the new procedure. The number of items in a set is denoted by x, and the data are in order of packing.

x	40	21	62	51	24	30	10	57	48	18	38	45
y	545	370	525	450	325	315	200	410	360	245	320	345

a Plot a scatter diagram of these data and label the points from 1 to 12 according to the order of packing.

b **i** Calculate the equation of the least squares regression line of y on x.
ii Draw the line on your scatter diagram.
iii Comment on the pattern revealed, and suggest why it has occurred.

c The least squares regression line for the last 6 points only is $y = 161 + 4.23x$. Draw this line on your scatter diagram.

d The packaging manager estimated that to deal with the next order, for 44 items of crockery, it would take 400 minutes.

i Comment on this estimate and the method by which you think it was made.

ii Make your own estimate of the time for dealing with this order and explain why you think it is better than that of the packaging manager.

5 A local amateur dramatic society wishes to estimate how much to spend on advertising for its production. The data below refer to the society's previous eight productions and show the spending on advertising and the subsequent number of tickets sold. (All values are recorded to the nearest 10.)

Advertising (£)	200	330	690	580	490	390	610	110
Tickets sold	120	210	400	290	310	250	370	130

a Identify, with a reason, the explanatory variable, x, and the response variable, y.

b Plot a scatter diagram of these data.

c **i** Calculate the equation of the least squares regression line in the form $y = a + bx$.
ii Draw the line on your scatter diagram.

d Interpret, in context, your value for b.

e **i** Calculate, to the nearest integer, the values of the eight residuals and confirm that their sum is zero.
ii Ignoring the signs of your residuals, calculate their average value.

f **i** Estimate, to the nearest 10, the number of tickets sold if the society decides to spend £500 on advertising.

   ii Using both your scatter diagram and your answer to **e ii**, comment on the likely accuracy of your estimate.

6 A road haulage contractor owns four lorries of the same specification and age. The contractor employs four drivers, A, B, C and D. For a number of long journeys, data is collected on the driver, weight of load carried, x kg, and the diesel consumption, y km/litre.

Driver	B	D	A	C	C	D	A	D	B	A
x	6600	6250	8300	7600	6950	5500	8450	7800	10 100	5650
y	6.11	5.77	5.62	5.89	5.99	5.88	5.49	5.15	4.98	6.22

a Plot a scatter diagram of these data, labelling the points according to the driver.

b **i** Calculate the equation of the least squares regression line of y on x.

   ii Draw the line on your scatter diagram.

c Interpret, in context, your values for the slope and for the intercept of the regression line.

d Comment on the diesel consumption of lorries driven by driver D.

e Why was the regression line of y on x calculated rather than the regression line of x on y?

f Given that 1016 kg equals 1 tonne and 1 km/litre equals 2.825 miles per gallon (mpg), find the equivalent relation to that found in **b i** relating diesel consumption in mpg to load in tonnes.

SKILLS CHECK **6D EXTRA** is on the CD

Examination practice Correlation and regression

1 A market trader sells ball-point pens on his stall. He sells the pens for a different fixed price, x pence, in each of six weeks. He notes the number of pens, y, that he sells in each of these six weeks. The results are shown in the following table.

x	10	15	20	25	30	35
y	68	60	55	48	38	32

Calculate the equation of the least squares regression line of y on x. [AQA(A) Jan 2004]

2 A biologist assumes that there is a linear relationship between the amount of fertilizer supplied to tomato plants and the subsequent yield of tomatoes obtained.

Eight tomato plants, of the same variety, were selected at random and treated, weekly, with a solution in which x grams of fertilizer was dissolved in a fixed quantity of water. The yield, y kilograms, of tomatoes was recorded.

Plant	A	B	C	D	E	F	G	H
x	1.0	1.5	2.0	2.5	3.0	3.5	4.0	4.5
y	3.9	4.4	5.8	6.6	7.0	7.1	7.3	7.7

a Plot a scatter diagram of yield, y, against amount of fertilizer, x.

b Calculate the equation of the least squares regression line of y on x.

c Estimate the yield of a plant treated, weekly, with 3.2 grams of fertilizer.

d Indicate why it may **not** be appropriate to use your equation to predict the yield of a plant treated, weekly, with 20 grams of fertilizer. [AQA(A) Jan 2002]

3 a Estimate, without undertaking any calculations, the value of the product moment correlation in each of the scatter diagrams below.

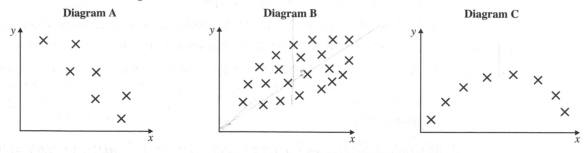

Diagram A   Diagram B   Diagram C

b The product moment correlation coefficient is an unsuitable measure of association for the data illustrated in one of the diagrams in part **a**. State, giving a reason, which diagram this is.

[AQA(B) Jan 2001]

 4 A mathematics teacher recorded the length of time, y minutes, taken to travel to school when leaving home x minutes after 7 am on seven selected mornings. The results are as follows.

x	0	10	20	30	40	50	60
y	16	27	28	39	39	48	51

a Plot the data on a scatter diagram.

b i Calculate the equation of the least squares regression line of y on x, writing your answer in the form $y = a + bx$.

ii Draw the regression line on your scatter diagram.

c The mathematics teacher needs to arrive at school no later than 8.40 am.

The number of minutes by which the mathematics teacher arrives early at school, when leaving home x minutes after 7 am, is denoted by z.

i Deduce that

$z = (100 - a) - (1 + b)x.$

ii Hence estimate, to the nearest minute, the latest time that the mathematics teacher can leave home without then arriving late at school. [AQA(A) June 2004]

5 As part of a biology project, the heights, in metres, and the weights, in kilograms, of all the students in a class were measured. The mean height was 1.71 and the standard deviation of the heights was 0.041. The product moment correlation coefficient between the heights and weights was 0.831.

If the heights had been measured in centimetres instead of metres, but the weights had still been measured in kilograms, write down the values of:

a the mean height;

b the standard deviation of the heights;

c the product moment correlation coefficient between the heights and weights. [AQA(B) May 2004]

 6 Naomi attends a health club. As part of her training programme, she is timed, on different occasions, to carry out predetermined numbers of step-ups. The table below shows the number of step-ups, x, and the time taken, y seconds, to complete them.

x	20	25	30	35	40	45	50	55	60	65
y	53	66	77	96	125	147	186	222	254	298

a Draw a scatter diagram of the data.

b Calculate the equation of the regression line of y on x and draw the line on your diagram.

c Calculate the residuals for the points where Naomi completed:

    i 45 step-ups;   **ii** 65 step-ups.

d **i** Use your regression equation to predict the time that Naomi would take to complete 62 step-ups.

    ii Give a reason why the prediction made in part **d i** is unlikely to be accurate.

e A new trainer suggests that, instead of undertaking a predetermined number of step-ups, Naomi should complete as many step-ups as possible in predetermined periods of time. How, if at all, would your method of calculation of the regression equation for data generated in this way differ from your method of calculation in part **b**? Justify your answer.   [AQA(B) Nov 2004]

7 The following table shows, for a sample of towns in England, the number of bank managers, x, living in the town and the total amount of unpaid council tax, y, in thousands of pounds.

x	3	45	8	23	32	17	12	9	24
y	102	1357	435	612	806	324	412	157	719

a Calculate the value of the product moment correlation coefficient between x and y.

b Using the context above as an example, explain whether a high correlation coefficient implies cause and effect.   [AQA(B) June 2003]

8 The following table shows the weekly gas and electricity consumption for a house in Manchester for a sample of nine weeks in 2001.

Week	1	2	3	4	5	6	7	8	9
Gas consumption, x kWh	312	46	23	406	350	67	295	247	110
Electricity consumption, y kWh	84	57	54	96	82	63	59	73	60

a Calculate the value of the product moment correlation coefficient.

b The householder expects that, for weeks when the gas consumption is high, the electricity consumption will be low and vice versa.

    i State to what extent the value you calculated in part **a** confirms or denies the householder's expectation.

    ii Give a reason why the value you calculated in part **a** is plausible.   [AQA(B) June 2002]

9 Megan and David run a small business installing double glazing. They give each job a score based on the size, number and accessibility of the windows to be double glazed. The price charged is related to this score.

The following table shows the score, x, given to each of nine jobs and the time, y hours, taken by Megan and David, working as a team, to complete each job.

Job	1	2	3	4	5	6	7	8	9
x	23	34	47	44	57	16	73	40	28
y	8.5	14.0	22.5	28.0	33.5	6.0	40.0	17.5	12.5

a Draw a scatter diagram of these data.

b Find the equation of the regression line of y on x and draw this line on your scatter diagram.

c Evaluate the residual for Job 4.

d The next job is given a score of 52.

    i Use your regression equation to estimate the time taken to complete this job.

    ii Also bearing in mind your result in part **c**, comment on Megan's claim that it is very unlikely that this job will take more than 34 hours.   [AQA(B) June 2002]

10 During train journeys, Ariane sells refreshments from a trolley to passengers at their seats. The following table shows for nine journeys the time, x minutes, she spent on the train and the value, £y, of her sales.

Journey	1	2	3	4	5	6	7	8	9
x	35	124	66	84	77	106	44	90	52
y	52	113	54	58	84	80	61	72	44

a Plot a scatter diagram of the data.

b Calculate the equation of the regression line and draw the line on your scatter diagram.

c The following table shows the residuals for some of the journeys.

Journey	1	2	3	4	5	6	7	8	9
Residual	7.6	15.1	−9.1	−15.9	14.3	−7.1	11.2		

    i Calculate the residuals for journeys 8 and 9.

    ii Find the standard deviation of the nine residuals.

d Desmond carries out the same duties as Ariane. His journey times and sales were recorded for nine journeys similar to those recorded for Ariane. The equation of the regression line for Desmond's journeys was $y = 50.2 + 0.290x$. The standard deviation of the residuals about this regression line was 6.2.

    i Draw Desmond's regression line on your scatter diagram.

    ii Compare Desmond's sales performance with that of Ariane.   [AQA(B) June 2004]

Practice exam paper B (without coursework)

Answer **all** questions.

Time allowed: 1 hour 30 minutes

Total marks: 75

You may use a graphics calculator and the **blue** AQA booklet of formulae and statistical tables.

1 The weight, in kilograms, of luggage belonging to each of a sample of 200 foot passengers on a ferry is summarised in the table.

Weight	0–	20–	30–	35–	40–	60–100
Number of passengers	31	40	37	51	22	19

    a Calculate estimates of the mean and standard deviation of these weights of luggage. *(4 marks)*

    b On another ferry, the mean weight of luggage, again for a random sample of 200 foot passengers, was 35.85 kg and the standard deviation was 10.64 kg.

    Compare, briefly, the weights of luggage belonging to foot passengers on the two ferries. *(2 marks)*

2 The proportion of a certain variety of seed that fail to germinate is 0.07.

    a A seedsman plants 40 such seeds. Find the probability that:

        i at most three seeds will fail to germinate;

        ii at least two seeds will fail to germinate;

        iii exactly five seeds will fail to germinate. *(6 marks)*

    b A householder plants 18 such seeds. Calculate the probability that exactly 15 of these seeds germinate. *(3 marks)*

    c State **one** assumption that you have made in answering parts **a** and **b**. *(1 mark)*

3 The height, in centimetres, of each girl, in a sample of nine sets of twin girls aged 16 years, was recorded with the following results.

Set of twin girls	A	B	C	D	E	F	G	H	I
Height of first born (x cm)	153	174	162	148	163	155	158	162	149
Height of second born (y cm)	149	165	163	146	160	148	153	156	185

a A scatter diagram of these data is shown.

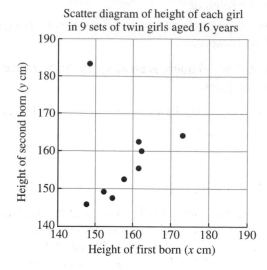

Scatter diagram of height of each girl
in 9 sets of twin girls aged 16 years

Give **two** distinct comments, in context, on what this diagram reveals. *(2 marks)*

b Subsequently, it was discovered that one of the nine sets of twins contained a girl and a boy, rather than two girls.

Identify which set of twins this is most likely to be. *(1 mark)*

c **Removing the data for the set of twins you identified in part b,** calculate the value of the product moment correlation coefficient for the remaining eight pairs of values of x and y. *(3 marks)*

4 Rea and Sally are members of a local flower club. The probability that Rea attends a club meeting is 0.96. The probability that Sally attends a club meeting is 0.70 when Rea attends the meeting, but is only 0.35 when Rea does not attend the meeting.

a Calculate the probability that

i neither attend a meeting; *(2 marks)*

ii exactly one of them attends a meeting. *(3 marks)*

b Tara is also a member of the flower club. The probability that she attends a meeting is 0.85 and this is not dependent on whether Rea or Sally attends the meeting.

Calculate the probability that

i none of the three members attend a meeting; *(2 marks)*

ii exactly one of the three members attends a meeting. *(3 marks)*

5 The times, to the nearest hour, spent during a particular week by a random sample of 40 teenagers on their computers were as follows.

20	9	15	28	6	0	2	12	32	10	41	27	0	8	17	6	0	22	33	27
13	21	0	17	11	0	16	31	13	0	17	0	21	12	0	27	35	26	37	0

The standard deviation of the time spent per week by teenagers on their computers is known to be 12 hours.

a i Calculate the mean of the above 40 times. *(1 mark)*

ii Give **two** reasons why it is unlikely that the above 40 times are normally distributed. *(2 marks)*

b **i** Indicate why the mean of the above 40 times may be assumed to be approximately normally distributed. *(2 marks)*

    ii Construct a 99% confidence interval for the mean time spent during the week by teenagers on their computers. *(5 marks)*

c One teenager's father suggests that 30 hours is an abnormally long time to spend on a computer during the week.

Comment on this suggestion using **both** the given information and your confidence interval. *(3 marks)*

6 [A sheet of graph paper is required for use in answering this question.]

A garden centre employs casual staff on Sundays. The manager carried out an investigation in order to see if there was a relationship between the number of casual staff employed, x, and the centre's Sunday takings, y thousands of pounds. The manager collected the following data from a random sample of 10 Sundays.

Sunday	A	B	C	D	E	F	G	H	I	J
x	4	6	7	9	11	13	14	16	17	19
y	5.1	6.4	7.1	7.8	6.5	8.8	9.4	12.4	11.6	11.9

a Plot a scatter diagram of these data. *(3 marks)*

b Calculate the equation of the least squares regression line of y on x and draw your line on your scatter diagram. *(6 marks)*

c Calculate the values of the residuals for Sundays **E** and **H**. *(3 marks)*

d It is known that, on one of the 10 Sundays, there was a nearby River Festival that attracted a large crowd.

Suggest, with a reason, on which Sunday this is most likely to have taken place. *(2 marks)*

7 A coach company operates a regular service between Leeds and Newcastle. The journey time may be modelled by a normal distribution with a mean of 165 minutes and a standard deviation of 10 minutes.

a Determine the probability that a particular journey time is:

    i less than 175 minutes; *(3 marks)*

    ii between 160 minutes and 180 minutes. *(3 marks)*

b As part of an advertising campaign, the company decides to offer a refund of fares when a journey time is more than m minutes.

Find the value of m such that the probability of fares being refunded for a particular journey is 0.002. *(4 marks)*

c Determine the probability that:

    i the mean time of 4 randomly selected journeys will be less than 160 minutes; *(4 marks)*

    ii the mean time will be less than 160 minutes for 4 randomly selected passengers from all those passengers on a particular journey. *(2 marks)*

Answers

SKILLS CHECK 1A (page 4)

1. 50, 2.08
2. 5.6, 1.64
3. 36, 138
4. **a** 15.2, 1.74
   b Times not independent; may be influenced by bad weather, road works, traffic volume, etc.
5. 2.47, 1.48
6. 6.01, 3.51, 1.87
7. 17.7, 3.13
8. 37.1, 6.56, 2.56
9. 201, 1.10, 1.21

SKILLS CHECK 1B (page 7)

1. 400, 1600
2. 50, 4
3. 1.01, 0.004
4. 5.31, 0.245
5. **a** 0.17, 0.646   **b** £3.31, £0.41

SKILLS CHECK 1C (page 11)

1. **a** 14, 14, 14.7   **b** 10, 4, 2.87
   c Mean and standard deviation as no 'unusual' values and involve all values in calculations
   Median and interquartile range as observable values
2. **a** 8, 9.5, 12   **b** 31, 6, 8.21
   c Mean and interquartile range as 2 'unusual' values
3. **a** 2.11, 2.12   **b** 0.22, 0.09, 0.0690
   c Mean and standard deviation as no 'unusual' values and involve all values in calculations
4. **a** 2.35, 2.56   **b** 2.79, 0.53, 0.757
   c Median and interquartile range an 'unusual' value
   d **i** Does not affect median and interquartile range as 2.84 is still the largest value
   Will reduce mean as total is reduced
   Will reduce range and standard deviation as maximum value is reduced
        ii Now use mean and standard deviation as no 'unusual' values and they both use all values in calculations
5. **a** 50, 50, 50.5   **b** 4, 1, 0.898
   c Mean and interquartile range as observable values or mean and standard deviation as they involve all values in calculations
6. **a** 20, 20, 20.3   **b** 6, 1, 1.15
   c Mean and standard deviation as involve all values in calculations or median and interquartile as observable values
7. **a** 2, 3.57   **b** 3, 4.90
   c Median and interquartile range as observable values and do not involve the upper class ranges

Exam practice 1 (page 12)

1. 2258, 104.7
2. 36.2, 9.92
3. **a** 0, 1, 1   **b** 4, 2, 1.10
   c Mean and standard deviation as no 'unusual' values and involve all values in calculations
   Median and interquartile range as observable values
4. **a** 52, 19   **b** 54.8, 18.2 to 18.5   **c** 54
   d **i** Median as not possible to calculate mean given missing values
        ii 54
5. **a** 6, 5, 6, 3
   b Median and interquartile range as mode = range = 6 is 'obvious'
   c Largest frequencies in tails

6. **a** 140.57 to 141.00, 22.39 to 22.47
   b **i** 147.59 to 148.05, 23.50 to 23.60; 147.79 to 148.25, 23.05 to 23.14
        ii Proposal 1 as resulting average salary is smaller
7. **a** 0, 15, 30   **b** 190, 45, 42.9 to 43.8
   c Median and interquartile range as many small values and one very large value
8. **a** 51.4, 14.9 to 15.1
   b **i** On average guests are much older and their ages are slightly less varied
        ii Children unlikely to be guests at a gourmet weekend
9. **a** 23, 2.98 to 3.03
   b For July 1900, $\bar{y} = 21.5\,°C$ and $s_y = 4.5\,°C$
   Temperatures in July 2000 were higher and less varied

SKILLS CHECK 2A (page 17)

1. **a** 0.45, 0.65   **b** 0.90, 0.10   **c** No; $P(A \cup B) \neq 1$
2. **a** 0.70, 0.45   **b** 0.85   **c** 0.15, 0.60
3. **a** 0.15   **b** 0.80
   c No; $P(A') + P(B') > 1$
4. **a** 0.65   **b** 0.40   **c** 0.60
5. **a** $\frac{1}{4}$ **b** $\frac{3}{4}$ **c** $\frac{3}{13}$ **d** $\frac{11}{26}$
   e $\frac{19}{52}$ **f** $\frac{5}{13}$ **g** $\frac{9}{13}$
6. **a** 0.24   **b** 0.80   **c** 0.64   **d** 0.60
   e 0.56   **f** 0.80   **g** 0.36
7. **a** **i** 0.60   **ii** 0.75   **iii** 0.25   **iv** 0.10
   b **i** C and D **ii** B, C and D
8. **a** **i** 0.20   **ii** 0.52   **iii** 0.588   **iv** 0.872   **v** 0.536
   b **i** Male, or female under 65 OR anyone not a female aged 65 and over
        ii Female aged 18 to 64
   c **i** C and R **ii** M and C, M and R

SKILLS CHECK 2B (page 21)

1. **a** 0.22, 0.044   **b** 0.12, 0.216
2. **a** 0.06   **b** 0.10
3. **a** 0.03   **b** 0.18   **c** 0.008   **d** 0.16
4. **a** $\frac{2}{15}$ **b** $\frac{1}{15}$ **c** $\frac{2}{9}$ **d** $\frac{1}{15}$
5. **a** $\frac{1}{4}$ **b** $\frac{3}{26}$ **c** $\frac{1}{13}$
6. **a** $\frac{25}{102}$ **b** $\frac{8}{221}$ **c** $\frac{8}{221}$
7. **a** 0.275   **b** 0.286   **c** 0.0195
   d 0.117   **e** 0.444   **f** 0.614
8. **a** 0.48   **b** 0.01   **c** 0.062   **d** 0.516

SKILLS CHECK 2C (page 24)

1. **b** **i** 0.0922   **ii** 0.0212
2. **a** 0.8245   **b** 0.171
3. **a** **i** 0.0425   **ii** 0.9575   **b** 0.992
4. **a** $\frac{1}{12}$ **b** $\frac{11}{120}$ **c** $\frac{7}{24}$ **d** $\frac{1}{24}$ **e** $\frac{1}{4}$
5. **a** 0.67   **b** 0.922
6. **a** **i** 0.72   **ii** 0.12   **iii** 0.28
        iv 0.18   **v** 0.84   **vi** 0.52
   b **i** 0.389   **ii** 0.737
7. **a** **i** 0.455   **ii** 0.15   **iii** 0.405   **iv** 0.30
   b 0.351
8. **a** **i** 0.375   **ii** 0.786   **iii** 0.826
   b **i** 0.339   **ii** 0.60
9. **a** **i** $\frac{1}{3}$ **ii** 0.08   **iii** 0.175   **iv** 0.564   **v** $\frac{1}{3}$
   b **i** C and S
        ii M and S; $P(M \cap S) \neq 0$, $P(S) = 0.08$, $P(S \mid M) = 0.064$
        iii M and C; $P(C) = 0.564$, $P(C \mid M) = 0.564$

Exam practice 2 (page 26)

1. **a** **ii** 0.3   **b** $\frac{2}{3}$
2. **a** **i** 0.005   **ii** 0.855   **iii** 0.140
   b 0.679

3 a i $\frac{13}{30}$ ii $\frac{2}{15}$ iii $\frac{1}{15}$ iv $\frac{7}{10}$ v $\frac{7}{15}$
   b i $\frac{2}{13}$
     ii No, $P(G) = \frac{2}{15}$, $P(G \mid B) = \frac{2}{13}$
4 a 0.68   b $\frac{14}{17}$
5 a i 0.48   ii 0.2   iii 0.68
   b $\frac{5}{17}$
6 a i $\frac{4}{15}$ ii $\frac{4}{15}$ iii $\frac{8}{15}$
   b $\frac{1}{2}$
7 a 0.33   b 0.26   c 0.53   d $\frac{26}{53}$
8 a i $\frac{3}{5}$ ii $\frac{9}{25}$ iii $\frac{29}{50}$ iv $\frac{3}{5}$
   b Yes; $P(M) = P(M \mid D) = \frac{3}{5}$
   c $\frac{9}{17}$
9 a 0.3   b 0.135
   c i 0.12   ii 0.3
   d 0.21
10 a i $\frac{19}{75}$ ii $\frac{3}{79}$
   b 0.162
   c i 0.127   ii 0.407   iii 0.762

SKILLS CHECK 3A (page 32)

1 a No, n is not fixed   b Yes, $n = 60$, $p = \frac{1}{6}$
2 a Yes, $n = 10$, $p = 0.10$
   b No, p is not constant **OR** trials are not independent
3 Trials may not be independent **OR** p is unlikely to be 0.65
4 p is almost constant as $10 \ll 6400$; $n = 10$, $p = 0.175$
5 a Fixed, independent (with replacement) trials with success (red) and failure (non-red) as outcomes; $n = 24$, $p = 0.40$
   b n is not fixed; p is not constant
6 a B(60, 0.65)
   b No, n is not fixed
   c p is likely to reduce over time due to learning effect

SKILLS CHECK 3B (page 34)

1 a 0.349   b 0.543   c 0.998   d 0.377
2 a 0.122   b 0.0384   c 0.546
3 a 0.736   b 0.191   c 0.838   d 0.632
4 a i 0.335   ii 0.402   iii 0.263
   b 0.296
   c 0.492
5 a i 0.242   ii 0.241   iii 0.271
   b i 0.0281
     ii A random sample **OR** independent for satellite TV
     iii Doubtful as p unlikely to be 0.36
6 a i 0.329   ii 0.351
   b 0.648
   c i 0.827   ii 0.181
7 a i 0.234   ii 0.221   iii 0.438   iv 0.508   v 0.489
   b 0.404
8 a i 0.286   ii 0.670
   b i 0.973   ii 0.712

SKILLS CHECK 3C (page 36)

1 a 0.2122   b 0.1148   c 0.0974
   d 0.7858   e 0.7878
2 a 0.7264   b 0.7199   c 0.0875   d 0.8022
3 a 0.9456   b 0.4161   c 0.9578   d 0.0002
4 a 0.0783   b 0.1763   c 0.2900   d 0.0971
5 a 0.9568   b 0.3231   c 0.2852   d 0.0769
6 a 0.2801   b 0.7138   c 0.0890
   d 0.0769   e 0.5836
7 a i 0.5610   ii 0.0978   iii 0.1399   iv 0.9507
   b i 0.4373   ii 0.8746   iii 0.1254
8 a i 0.4164   ii 0.0480   iii 0.1789
   b ii 0.3385

SKILLS CHECK 3D (page 39)

1 17.6, 11.3, 3.36
2 a 0.2
   b i 6, 2.19   ii 0.1795
3 a 20, 3.16
   b i 0.1254
     ii 0.9616
     iii Yes; for B(40, 0.5) would expect about 95%
4 a 4, 3.2
   b i 4, 3.2 or 3.28
     ii (Almost) exact agreement, suggesting Jamil has fiddled his results!
5 a 3, 1.31
   b i 3, 2.83
     ii Doubtful as 'large' discrepancy in standard deviations
6 a i 0.3, 0.577   ii 0.1
   b i 0.3, 0.520
     ii 0.7290, 0.2430. 0.0270, 0.0001
     iii 73, 24, 3, 0
   c Claim appears reasonable as close agreement in standard deviations and in each pair of values for number of samples

Exam practice 3 (page 40)

1 a 0.2894   b 0.2011   c 0.1465
2 a 0.0644   b 0.9956   c 0.0190
3 a i 0.9790   ii 0.2645
   b 0.1954
4 a 0.1642   b 0.7421   c 0.648
5 a 0.1172
   b 0.7553
   c Number of trials, n, is not fixed.
6 a i 0.5886   ii 0.155
   b Random samples
   c P(blue) is not constant as number of blue beads on each string is unlikely to be the same.
7 a i 0.0479   ii 0.4711   iii 0.6926
   b Number of trials, n, is not fixed.
     Probability of yellow eraser, p, is not constant.
8 a i 0.5518   ii 0.2965   iii 0.1941
   b No, n not fixed, trials not independent, 0 and 1 are not possible outcomes.
   c No, p not constant, trials not independent.
9 a i 0.9734   ii 0.0022   iii 0.75, 0.798
   b i 0.75, 1.24 or 1.23
     ii 0.15
   c i Not plausible; standard deviation much larger
     ii Some pupils more likely to be late than others
     Late arrivals unlikely to be independent

SKILLS CHECK 4A (page 47)

1 a 0   b 0.80511   c 0.08851
   d 0.62522   e 0.10204
2 a 0.16287   b 0.48030   c 0.74628   d 0.39069
3 a 0.78814   b 0.05480   c 0
   d 0.65542   e 0.11507
4 a 0.37617   b 0.66124   c 0.34134   d 0.20114
5 a 0.90824   b 0.97725   c 0.53754   d 0.86638
6 a 0.61026   b 0.91774   c 0.40699
   d 0.30748   e 0   f 0.5
7 a i 0.84134   ii 0.33448
   b i 0.15866   ii 0.14644
8 a i 0.95254   ii 0.95254   iii 0.88549
   b 0.02275
9 a i 0.02619   ii 0.20327   iii 0.50899
   b Mark; $P(\text{jump} > 8) = 0.04272 > 0.02619$
10 a i 0.25143   ii 0.02275   iii 0.98631   iv 0.90812
   b 210

SKILLS CHECK 4B (page 51)

1 a 0.6745 **b** 1.9600 **c** 0.4399
d -2.0537 **e** -1.0364
2 a 0.9542 **b** 2.1701 **c** 0.6745
3 a 22.1 **b** 21.7 **c** 18.7 **d** 17.4
4 a 140.2, 165.8 **b** 147.8, 158.2
5 a i 32.6 **ii** 26.7
b 26.1, 33.9
6 224
7 a 0.0860 **b** 0.03
8 1018, 27.8
9 a i 748 **ii** 474
b i 9.5%
ii 910
10 a i 4.75 **ii** 90.5
b 0.007
c 9.508, 0.007

Exam practice 4 (page 53)

1 a 0.15866 **b** 34
2 a 0.68268 **b** 150.6
3 a i 0.0548 **ii** 0.645 **b** 25.46
4 a 0.19766 **b** 67.5
5 a 0.10565 **b** 0.122
6 a 0.15866 **b** 128.4 **c** 1.31
7 a i 0.833 **ii** 33.4 **b** 2.4
c $X: P(X < 35) = 0.9$, $P(Y < 35) = 0.8$
8 a i 0.841 **ii** 154.5 **b** 181, 10
9 a 0.841 **b** 2 : 41 pm **c** B; smaller standard deviation
10 a i 0.894 **ii** 0.493
b Longest possible stay is 60 minutes, but for proposed model about 60% of times will exceed this value
c 6.55 pm (using $\mu + 3\delta$)
11 a 0.785 **b** 327 **c** 216
d i 5.94
ii Standard deviation of six seconds suggests an unreasonably small variation given differing lengths of queue and customer requirements.

SKILLS CHECK 5A (page 58)

1 a i 0.106 **ii** 0.702
b i 0.174 **ii** 0.394
2 a 0.346 **b** 0.0749
3 a 0.794 **b** 0.977
c Answer to **a** perhaps not valid as $6 < 30$ so CLT does not apply
Answer to **b** remains valid as $36 > 30$ so CLT does apply
4 a Mean $+ 1 \times$ standard deviation $> 100\%$ which is impossible
b i N(83, 8.82) using CLT as $50 > 30$
ii 0.592
5 a Mean $- 2 \times$ standard deviation < 0 which is impossible
b 0.733
6 a 0.0808 **b i** N(25.8, 0.005104) **ii** 0.997

SKILLS CHECK 5B (page 60)

1 a i (250, 266) **ii** (246, 270) **b** (255, 261)
c i Increases width of CI
ii Decreases width of CI
2 (15.4, 19.8)
3 a (8.76, 10.9) **b** Claim likely to be valid as $11 > $ UCL
4 a (252, 258)
b CI shows mean > 250 so claim likely to be valid
Sample shows $25\% < 250$ so claim unlikely to be valid
Overall, claim unlikely to be valid
5 a (1.42, 1.92) **b** Claim likely to be valid as $2 > $ CI
6 a (0.97, 1.21)
b Claim unlikely to be valid as 1 metre is contained within CI

SKILLS CHECK 5C (page 63)

1 (135.3, 138.7)
2 a Mean $- 2 \times$ standard deviation $= 30$ which is unrealistic
b (76.5, 88.7)
3 a Mean $- 2 \times$ standard deviation < 0 which is impossible
b (21.75, 30.95)
c $2.5 \times 11.84 = 29.60$ which is within CI so claim likely to be valid
4 a (30.4, 35.2)
b Claim unlikely to be valid as UCL > 35
5 a i (29.7, 33.5)
ii Suspicion likely to be valid as $28.5 < $ LCL
iii Application of CLT as $65 > 30$
b 28.5
c i 43.5 **ii** 12.5
6 a 34.3, 27.04
b i (32.4, 36.2)
ii (20.9, 47.7)
iii From **i**, average speed > 30
From **ii**, many speeds < 30
Limit not being ignored altogether
c Answer to **i** still valid by CLT as $50 > 30$
Answer to **ii** not valid
Answer to **iii** restricted to average speed

Exam practice 5 (page 65)

1 0.933
2 a 0.960
b Central Limit Theorem, sample size is large
3 a 0.945
b Random sample or pins selected independently
4 (355.2, 357.8)
5 a 0.894 **b** Volume is normally distributed
6 a i 0.894 **ii** 10.2
b i 24.9 **ii** 0.02 **iii** (24.86, 24.94)
iv Disagree as 25 is outside CI
7 a 0.0082 **b** 49.7
c i (52.32, 52.60)
ii Central Limit Theorem as sample size of 36 is sufficiently large
8 a (454.6, 458.9)
b CI suggests mean > 454, sample shows not all jars have contents greater than 454.
c 0.05
9 a (311, 943)
b i Three people drank no water, suggesting mode of zero
ii Cannot be negative
c i (926, 994)
ii Large sample so mean approximately normal
iii Night shift/first day unlikely to be typical

SKILLS CHECK 6A (page 69)

1 0.554
2 0.805
3 0.954
4 -0.195
5 -0.939
6 0.150

SKILLS CHECK 6B (page 73)

1 a -0.100 to 0.100 **b** -0.800 to -0.980
c -0.400 to -0.800 **d** 0.400 to 0.800
2 a -0.741
b As number of items completed increases, there is some indication that the mean quality score decreases
c ii Correlation probably solely due to unusual result (Seamstress G)
d Seamstress G should be removed from analysis with result that there is no apparent correlation

3 a i 0.870

 ii As chest measurement increases, there is a strong indication that waist measurement also increases

 c i D, E, H, I, J, L

 ii Small waists relative to smaller chests

 d Analyse data for males and females separately

4 b 0.970

 c Diagram suggests very strong positive linear relationship

 d For each engine, results for Method B are above those for Method A

5 a i -0.234

 ii Little or no linear relationship between IQ and DI

 b ii Absence of correlation probably solely due to unusual result (teenager G)

 c i Values for teenager G will now match those for other teenagers

 ii -0.800 to -0.950

6 b i May be a non-linear relationship

 ii May be spurious correlation

7 a Plausible; longer tails on larger mice that are probably heavier

 b Definitely incorrect; outside range of -1 to $+1$

 c Probably incorrect; larger diameters imply greater weights

 d Spurious; use of toothpaste unlikely to cause fillings

 e Plausible; hotter weather suggests fewer hot drink sales

SKILLS CHECK 6C (page 76)

1 a $y = 4 + x$ **b** $y = 52 - 4x$

2 $y = 0.865 + 0.355x$

3 $d = 9.16 + 4.64\theta$

4 $y = 17.7 + 1.05x$

5 Deflection $= 0.946 + 0.0721 \times$ temperature

6 Length $= 93.15 + 0.595 \times$ weight

SKILLS CHECK 6D (page 81)

1 b i $y = 115 + 4.8x$

 c a is yield of process with no additive

 b is (average) increase in yield for each percentage increase in additive

 d i 135; likely to be reasonably accurate as interpolation

 ii 187; unlikely to be accurate as (large) extrapolation

2 b i $y = 6.88 - 0.00818x$

 c a is pH of skimmed milk at $0\,°C$

 b is (average) decrease in pH for each $1\,°C$ increase in temperature

 d i 6.57; likely to be reasonably accurate as interpolation

 ii 6.27; unlikely to be accurate as extrapolation

3 a Depth is x and sand content is y as latter depends on former

 c i $y = 105 - 1.03x$

 d Decrease in percentage sand content for each 1 cm increase in depth

 e 105; nonsense as it cannot be greater than 100

4 b i $y = 195 + 4.66x$

 iii For similar values of x, later values of y are smaller than earlier values due to 'learning' effect

 d i Appears high due to use of equation in **b i**

 ii 347 using equation in **c** which takes account of recent 'learning' effect

5 a Advertising is x and tickets sold is y as latter depends on former

 c i $y = 52.6 + 0.488x$

 d For each £100 increase in advertising, ticket sales increase by (on average) about 50

 e i $-30, -4, 11, -46, 18, 7, 20, 24$ **ii** 20

 f i 300

 ii Reasonably accurate as points show linear relationship and average deviation from line is relatively small

6 b i $y = 7.40 - 0.000231x$

 c b; for every 1000 kg increase in load carried, diesel consumption decreases by (on average) 0.231 km/litre

 a; diesel consumption with no load

 d An economic driver as all his points below line/other points

 e Diesel consumption depends upon load not the reverse

 f $v = 20.9 - 0.663u$

Exam practice 6 (page 83)

1 $y = 37.4 - 1.73x$

2 b $y = 3.25 + 1.08x$

 c 6.7

 d Outside data range

3 a A: -0.6 to -0.98 B: 0.1 to 0.6 C: -0.2 to 0.2

 b C as non-linear

4 b i $y = 18.5 + 0.564x$

 c ii 7.52 am

5 a 171

 b 4.1

 c 0.831

6 b $y = -81.4 + 5.50x$

 c i -19.1 **ii** 21.9

 d i 260

 ii Graph and residuals suggest that, in this region, actual > predicted

 e $x = a + by$ as step-ups depend upon time

7 a 0.956

 b Suggests towns with large numbers of bank managers have large amounts of unpaid council tax; No cause and effect; Large towns will have large numbers of both

8 a 0.850

 b i Inconsistent with expectations as high gas associated with high electricity

 ii Cold weather likely to lead to high consumptions of both gas and electricity, hot weather to low consumptions

9 b $y = -5.53 + 0.642x$

 c 5.28

 d i 27.8

 ii Claim plausible as 34 is 6.2 above 27.8 and largest residual is 5.3

10 b $y = 23.4 + 0.601x$

 c i $-5.5, -10.6$ **ii** 12.0 or 11.3

 d ii Ariane's sales lower on short journeys but higher on long journeys. Desmond's sales more predictable

Practice exam paper B (page 87)

1 a 35.25, 18.40 (or 18.35)

 b Average weights almost identical, much smaller spread of weights on second ferry

2 a i 0.694   **ii** 0.780   **iii** 0.0873

 b 0.0942

 c Seeds germinate independently

3 a Evidence of positive linear relationship between twins' heights

 One unusual pair of twins' heights

 b I

 c 0.918

4 a i 0.026   **ii** 0.302

 b i 0.0039   **ii** 0.0674

5 a i 15.3

 ii Modal time is zero

 Mean $-$ standard deviation $= 3$ so negative times likely

 b i Sample large (>30) so can apply Central Limit Theorem

 ii (10.4, 20.2)

 c CI is well above 30 but 6 in 40 (15%) of times above 30

 Suggestion is not valid

6 b $y = 3.18 + 0.467x$

 c E: -1.82 H: 1.80

 d E; greatest negative residual

7 a i 0.841   **ii** 0.625

 b 194

 c i 0.159   **ii** 0.309

WARNING: BY OPENING THE PACKAGE YOU AGREE TO BE BOUND BY THE TERMS OF THE LICENCE AGREEMENT BELOW.

This is a legally binding agreement between You (the user or purchaser) and Pearson Education Limited. By retaining this licence, any software media or accompanying written materials or carrying out any of the permitted activities You agree to be bound by the terms of the licence agreement below.

If You do not agree to these terms then promptly return the entire publication (this licence and all software, written materials, packaging and any other components received with it) with Your sales receipt to Your supplier for a full refund.

YOU ARE PERMITTED TO:

- Use (load into temporary memory or permanent storage) a single copy of the software on only one computer at a time. If this computer is linked to a network then the software may only be used in a manner such that it is not accessible to other machines on the network.

- Transfer the software from one computer to another provided that you only use it on one computer at a time.

- Print a single copy of any PDF file from the CD-ROM for the sole use of the user.

YOU MAY NOT:

- Rent or lease the software or any part of the publication.

- Copy any part of the documentation, except where specifically indicated otherwise.

- Make copies of the software, other than for backup purposes.

- Reverse engineer, decompile or disassemble the software.

- Use the software on more than one computer at a time.

- Install the software on any networked computer in a way that could allow access to it from more than one machine on the network.

- Use the software in any way not specified above without the prior written consent of Pearson Education Limited.

- Print off multiple copies of any PDF file.

ONE COPY ONLY

This licence is for a single user copy of the software

PEARSON EDUCATION LIMITED RESERVES THE RIGHT TO TERMINATE THIS LICENCE BY WRITTEN NOTICE AND TO TAKE ACTION TO RECOVER ANY DAMAGES SUFFERED BY PEARSON EDUCATION LIMITED IF YOU BREACH ANY PROVISION OF THIS AGREEMENT.

Pearson Education Limited and/or its licensors own the software.
You only own the disk on which the software is supplied.

Pearson Education Limited warrants that the diskette or CD-ROM on which the software is supplied is free from defects in materials and workmanship under normal use for ninety (90) days from the date You receive it. This warranty is limited to You and is not transferable. Pearson Education Limited does not warrant that the functions of the software meet Your requirements or that the media is compatible with any computer system on which it is used or that the operation of the software will be unlimited or error free.

You assume responsibility for selecting the software to achieve Your intended results and for the installation of, the use of and the results obtained from the software. The entire liability of Pearson Education Limited and its suppliers and your only remedy shall be replacement free of charge of the components that do not meet this warranty.

This limited warranty is void if any damage has resulted from accident, abuse, misapplication, service or modification by someone other than Pearson Education Limited. In no event shall Pearson Education Limited or its suppliers be liable for any damages whatsoever arising out of installation of the software, even if advised of the possibility of such damages. Pearson Education Limited will not be liable for any loss or damage of any nature suffered by any party as a result of reliance upon or reproduction of or any errors in the content of the publication.

Pearson Education Limited does not limit its liability for death or personal injury caused by its negligence.

This licence agreement shall be governed by and interpreted and construed in accordance with English law.